90 0819575 1

Nanotechnology, Risk and Communication

Also by Stuart Allan

MEDIA, RISK AND SCIENCE

NEWS CULTURE

JOURNALISM: Critical Issues (*edited*)

ONLINE NEWS: Journalism and the Internet

ENVIRONMENTAL RISKS AND THE MEDIA (*co-edited*)

Also by Alison Anderson

MEDIA, CULTURE AND THE ENVIRONMENT

MEDIA, ENVIRONMENT AND THE NETWORK SOCIETY

THE CHANGING CONSUMER: Markets and Meanings (*co-edited*)

Also by Alan Petersen

BIOBANKS: Governance in Comparative Perspective (*co-edited*)

GENETIC GOVERNANCE: Health, Risk and Ethics in the Biotech Era (*co-edited*)

THE BODY IN QUESTION: A Socio-Cultural Approach

THE NEW GENETICS AND THE PUBLIC'S HEALTH (*with R. Bunton*)

HEALTH, RISK AND VULNERABILITY (*co-edited*)

Nanotechnology, Risk and Communication

Alison Anderson
University of Plymouth, UK

Alan Petersen
Monash University, Australia

Clare Wilkinson
University of West of England, Bristol, UK

and

Stuart Allan
Bournemouth University, UK

© Alison Anderson, Alan Petersen, Clare Wilkinson and Stuart Allan 2009

All rights reserved. No reproduction, copy or transmission of this publication may be made without written permission.

No portion of this publication may be reproduced, copied or transmitted save with written permission or in accordance with the provisions of the Copyright, Designs and Patents Act 1988, or under the terms of any licence permitting limited copying issued by the Copyright Licensing Agency, Saffron House, 6-10 Kirby Street, London EC1N 8TS.

Any person who does any unauthorized act in relation to this publication may be liable to criminal prosecution and civil claims for damages.

The authors have asserted their rights to be identified as the authors of this work in accordance with the Copyright, Designs and Patents Act 1988.

First published 2009 by
PALGRAVE MACMILLAN

Palgrave Macmillan in the UK is an imprint of Macmillan Publishers Limited, registered in England, company number 785998, of Houndmills, Basingstoke, Hampshire RG21 6XS.

Palgrave Macmillan in the US is a division of St Martin's Press LLC, 175 Fifth Avenue, New York, NY 10010.

Palgrave Macmillan is the global academic imprint of the above companies and has companies and representatives throughout the world.

Palgrave® and Macmillan® are registered trademarks in the United States, the United Kingdom, Europe and other countries.

ISBN-13: 978–0–230–50693–0 hardback
ISBN-10: 0–230–50693–3 hardback

This book is printed on paper suitable for recycling and made from fully managed and sustained forest sources. Logging, pulping and manufacturing processes are expected to conform to the environmental regulations of the country of origin.

A catalogue record for this book is available from the British Library.

Library of Congress Cataloging-in-Publication Data
Nanotechnology, risk, and communication / Alison G. Anderson . . . [et al.].
    p. cm.
  Includes bibliographical references and index.
  ISBN 978–0–230–50693–0
  1. Nanotechnology.   2. Communication in science.   I. Anderson, Alison, 1965–
  T174.7.N375245 2009
  303.48′3—dc22

                                                                    2008046437

10 9 8 7 6 5 4 3 2 1
18 17 16 15 14 13 12 11 10 09

Printed and bound in Great Britain by
CPI Antony Rowe, Chippenham and Eastbourne

Contents

List of Tables

Acknowledgements

The support from the Economic and Social Research Council (ESRC) and the British Academy (BA) is gratefully acknowledged – ESRC: RES-000-22-0596; BA: SG-44284.

This book would not have been possible without the input and support of many people to whom we offer grateful thanks. We thank Rachel Torr for her dedicated and efficient research assistance on the British Academy study and Richard Handy for his invaluable assistance in gaining access to scientists working in the nanotechnologies field. Alison, Alan and Stuart are grateful to Clare Wilkinson, who was the researcher on the ESRC project, and who contributed to the development of ideas and the resulting publications. We are indebted to the scientists, policymakers and journalists who took part in our research and gave up their time to be interviewed.

Some parts of this book draw upon work that was published in different form elsewhere. We have drawn freely from:

Anderson, A., Allan, S., (2005), Petersen, A., and C. Wilkinson 'The Framing of Nanotechnologies in the British Newspaper Press', *Science Communication*, 27 (2), 200–220.

Petersen, A., Anderson, A., Allan, S., and C. Wilkinson (2008) 'Opening the Black Box: Scientists' Views on the Role of the Mass Media in the Nanotechnology Debate', *Public Understanding of Science*.

Petersen, A. and A. Anderson (2008) 'A Question of Balance or Blind Faith? Scientists' and Policymakers' Representations of the Benefits and Risks of Nanotechnologies', *NanoEthics*, 1 (3), 243–256.

Wilkinson, C., Allan, S., Anderson, A., and A. Petersen (2007) 'From Uncertainty to Risk?: Scientific and News Media Portrayals of Nanoparticle Safety', *Health, Risk and Society*, 9 (2), 145–157.

We thank the editors and publishers of *NanoEthics* for their permission to use material drawn from an earlier article.

1
Introduction

Nanotechnology is set to disrupt the face of much of industry. Nanotechnology is about new ways of making things. It
promises more for less: smaller, cheaper, lighter and faster
devices with greater functionality, using less raw material
and consuming less energy. Any industry that fails to investigate the potential of nanotechnology, and to put in place its
own strategy for dealing with it, is putting its business at risk.

(DTI/OST, 2002, 6)

Precisely what counts as 'nanotechnology' eludes easy explanation.
This may seem somewhat surprising to say, given the rapidly growing
number of references to it in different media contexts. Looking across
a range of these contexts – such as advertisements for exciting new
nano-products, certain (often dystopian) visions in science-fiction
cinema, novels and comic books, or even more nuanced representations in the science pages of a newspaper – is likely to reveal a
number of competing definitions. At stake, it seems, is more than
the usual sorts of disagreements over terminology and classifications
among scientists. That is to say, it would appear that what counts as
nanotechnology is also a problem of communication – and therefore,
quite possibly, one of risk where public perceptions are concerned.

To observe that nanotechnology revolves around the design and
manipulation of matter at the atomic and molecular level is to
acknowledge, at the same time, that its application invites a host
of questions about the likely implications for the way we live our
lives. Hailed by its proponents as the next Industrial Revolution,

nanotechnology has already attracted significant controversy over an apparent lack of adequate regulatory control (Michelson and Rejeski, 2006). It is neither a new nor a single technology; instead, it involves a fusion of elements of chemistry, physics, materials science and biology (Wood et al., 2007). Hence the term 'nanotechnologies' is sometimes adopted in order to capture this complexity. Confusing matters further, however, is the way nanotechnology cuts across a variety of different scientific fields, thereby necessitating a certain degree of interdisciplinarity. Moreover, older technologies are increasingly being repackaged as 'nano' in the fierce competition to attract funding. These and related issues, taken together, pose a major challenge for ethics and governance, particularly concerning how information about technological innovation is communicated during the early phases of development.

This book examines the increasingly crucial role played by the news media in framing the debate about potential benefits and risks of nanotechnologies. We argue that the ways in which the significance of possible risks associated with nanotechnology are recurrently framed in the early stages of their rising public visibility is likely to be a key factor in how citizens comprehend and subsequently respond to the technologies. This is especially important, we will suggest, with respect to whether they perceive the benefits as outweighing the risks, a social process of negotiation – economic, political and cultural – certain to impact upon levels of public trust to a significant extent.

Defining nanotechnology

The concept of 'nanoscience' was first described by physicist Richard Feynman in his lecture to the American Physical Society in 1959. This set out his vision of molecular manufacturing – factories using tiny machines at the molecular scale, designed and built atom by atom (see Drexler, 2004). The term 'nanotechnology' was introduced in 1974 by a Japanese scientist, Norio Taniguchi, to refer to precision engineering with tolerances of a micron or less (see Park, 2007). However, there are a whole variety of different definitions currently in circulation and a great deal of definitional ambiguity surrounds the label 'nanotechnology'. These range from the celebratory (which see it as a major breakthrough) to the dismissive (claims that it is nothing new, just applied chemistry with a fancy name to attract funding). As the Centre for Responsible Nanotechnology notes:

> Unfortunately, conflicting definitions of nanotechnology and blurry distinctions between significantly different fields have complicated the effort to understand the differences and develop sensible, effective policy.

(CRN, 2008)

According to Eric Drexler, there are two overarching definitions of nanotechnology: Feynman's original definition, based upon his vision of a molecular manufacturing system, and a more recent broader definition, referring to a host of products built at the nanoscale:

> In recent years a group of scientists, technologists, business leaders, and bureaucrats have exploited the excitement around nanotechnology by using the term to label existing and near-term products which have significant features less than 100 nanometers in size. By this new, loose definition, "nanotechnology" isn't about making nanoscale productive systems, but about making nanoscale products. It can describe anything with small features, ranging from fine particles to thin coatings to large molecules – even big things with tiny holes. Many parts of chemistry, materials science, microelectronics, and biotechnology are now marketed as 'nanotechnology'. This redefinition has created confusion, raised false expectations, and hampered progress toward the original, more powerful goal.

(Drexler, 2008)

The US National Science Foundation (NSF) defines nanotechnology as:

> Research and technology development at the atomic, molecular or macromolecular levels, in the length scale of approximately 1–100 nanometer range, to provide a fundamental understanding of phenomena and materials at the nanoscale and to create and use structures, devices and systems that have novel properties and functions because of their small and/or intermediate size. The

> novel and differentiating properties and functions are developed
> at a critical length scale of matter typically under 100 nm.
>
> (NSF, 2008)

Similarly, the Richard Smalley Institute for Nanoscale Science and
Technology describes it as:

> an emerging and promising field of research, loosely defined as
> the study of functional structures with dimensions in the 1–1000
> nanometer range. Certainly, many organic chemists have designed
> and fabricated such structures for decades via chemical synthe-
> sis. During the last decade, however, developments in the areas
> of surface microscopy, silicon fabrication, biochemistry, physical
> chemistry, and computational engineering have converged to pro-
> vide remarkable capabilities for understanding, fabricating and
> manipulating structures at the atomic level.
>
> (Smalley, 2008)

To give an indication of scale, one nanometre is one-billionth of a
metre, which is tens of thousands of times smaller than the diam-
eter of a human hair. Nanotechnology, as a distinctive field in its
own right, is still in its infancy, largely characterised by the application of methods from nanoscience to develop products (Wood et al.,
2003). There has been a considerable amount of re-branding of existing research in order to label it as 'nano' in ways that are likely to be
more attractive to funders. Also, it is difficult to abstract it from other
developments, such as information technology and biotechnology,
where there is increasing convergence and synergy (see Wood et al.,
2007). Thus where we use the generic term 'nanotechnology' in this
book it is important to recognise that it is interdisciplinary. Our usage
thus refers to a whole range of nanotechnologies, including a diverse
number of current or potential applications.

While some see nanotechnology as heralding the next Industrial Revolution (Theodore and Kunz, 2005), others argue that it
may engender 'horrendous social and environmental risks' (ETC
Group, 2003). Proponents of nanotechnology often argue that it has
potential far-reaching economic, health, environmental and other
benefits. According to some radical utopian visions it will eliminate problems of inequality, starvation and scarcity, thereby bringing

about a less-polluted environment [Dunkley (2004) cited in Wood et al., 2008, 15]. However, it has already evoked considerable controversy as a potentially dangerous development, with comparisons being made to previous disasters such as thalidomide and asbestos.

By way of an example, in March 2006 German newspapers reported that Magic-Nano aerosol products had led around 100 citizens to suffer illness with symptoms such as breathing difficulties, coughing, vomiting and headaches. A small number were treated in hospital for pulmonary oedema. However, it later transpired that this nanoscare was a false alarm; the problem was to do with the pH levels in the aerosol, and nanoparticles were not actually present in the product. Nonetheless, this incident led to renewed calls to investigate the risks associated with nanotechnology and to develop regulatory measures. Further concerns have been raised about Samsung's 'Nano Silver' washing machine, which was temporarily withdrawn from sale in Sweden in 2005, following public protests and concern from government regulators. Nano Silver is judged by some to pose unacceptable risks to beneficial bacteria in environmental systems and to human health. Indeed, Friends of the Earth Australia (FOEA) have called on the company to withdraw the range from sale in Australia until peer-reviewed studies can demonstrate its safety for the environment and human health. The environmental group also claim that there is accumulating evidence that carbon nanotubes could be the next asbestos, citing two recent studies published in *Nature Nanotechnology* and the *Journal of Toxicological Sciences* which found that multi-walled carbon nanotubes cause asbestos-like disease in mice (FOEA, 2008b).

Expenditure on research and development of nanotechnologies has dramatically accelerated in recent years and it promises to be one of the most rapidly growing areas of scientific innovation of the twenty-first century. Indeed, over the last ten years, it has become the fastest rising sector in the knowledge-based economic infrastructures in OECD countries (see Throne-Holst and Stø, 2008). As applied to medicine, nanoscience is seen to bring a broad range of benefits, including the development of artificial organs and implants, improved drug delivery, the cleaning of arteries, the repairing of cells and the diagnosis of disease (see Oud and Malsch, 2003; Wood et al., 2003, 21). Nanotechnology also has, or is predicted to have, applications in a range of other areas, such as sunscreens, self-cleaning windows, nanocrystalline alloys, micromachined silicon sensors (used in cars, cameras, etc.), films and coatings, fuel cells and bioremediation (systems capable of fixing heavy metals, PCBs, cyanide and other environmentally damaging materials) (Royal Society and Royal Academy of Engineering [RS/RAE], 2004). It is thought that the market for such products could rise to trillions of Euros in the coming years (see Cordis, 2004).

Our investigation

This book sets out to examine how the benefits and risks associated with nanotechnology have been framed by the news media and within the wider policy-making arena. Our particular interest is in tracing the framing of nanotechnologies in the context of prior controversies over emerging technologies, such as genetically modified (GM) food and crops, embryonic stem cells and human cloning. This includes a consideration of how the parameters of the debate have been represented within policy documents. 'Communication', then, is used in a broad sense to encompass more than simply considering the role of the news media. In addition to examining media portrayals we explore scientists' and policymakers' representations of nanotechnology, since previous studies have demonstrated that they may play a decisive role in shaping policy outcomes through influencing media coverage (e.g. Nisbet and Lewenstein, 2002). We also examine how 'the public' is conceptualised by experts working in this field and explore the question of what 'public engagement' means in practice. The 'public' is often left unproblematised in official discussions, which often employ simplistic formulations of 'public opinion'. Where we refer to the 'public', it is important to note that there are multiple publics with diverse views on science and technology.

In addition to offering a synthesis of the pertinent research literature in the media, science, technology and risk area of enquiry, the book draws on data from one of the first Economic and Social Research Council (ESRC)-funded projects on nanotechnology in the United Kingdom (UK), conducted between 2003 and 2005 and involving all four authors (Anderson, Petersen and Allan with Wilkinson as the then research associate). This brings together the key findings from our quantitative content analysis of UK national press coverage and in-depth, qualitative interviews with scientists, journalists and editors to ascertain their views on the production and coverage of news on nanotechnology. These findings are supplemented by those from further investigations conducted by the authors separately or in smaller groups. In the case of the latter, a British Academy-funded study conducted by Petersen and Anderson has added an important dimension to our discussion here. It examined how scientists and science policymakers, working in the field of nanotechnologies, seek to strike a balance in representing their benefits and risks.

It is widely recognised in the social sciences that the levels of attention devoted to a problem are not an accurate reflection of its 'objective' seriousness. Studies of previous biotechnology controversies highlight the socially constructed nature of news coverage and the influence of policy arenas and cultural factors (see Hornig Priest, 2001; Nisbet and Huge, 2006, 2007; Nisbet and Lewenstein, 2002). As Hilgartner and Bosk (1988) observe, issues compete with one another for attention and there is often considerable competition among news sources vying for position, each of whom is likely to be actively seeking to establish their definition of the issues in question as the preferred one to be adopted. Past studies suggest that initial press coverage of biotechnology was relatively sparse and overwhelmingly positive in tone (see Bauer and Gaskell, 2002; Nisbet and Huge, 2006). As we will detail later on in this book, early nanotechnology coverage follows a similar pattern. In the case of past controversies over embryonic stem cell research in the United States (US), it was only when the issue became high on the political agenda that media coverage peaked (see Nisbet et al., 2003). Thus far national press coverage of nanotechnology in the UK has been minimal and largely restricted to elite newspapers; short bursts of more widespread attention can be related to specific policy events and interventions by high-profile actors – in this particular case Prince Charles (see Chapter 4). The issues have yet to move beyond the administrative context and come to be seen as of pressing significance in overtly political arenas.

In the case of GM food and crops in the UK, following in the wake of the food scare surrounding Bovine Spongy Encephalopathy (BSE), particularly potent metaphors emerged as the issues became a political 'hot potato' (see Rowell, 2003). A string of sensational headlines appeared in the national press including: 'Alarm over

"Frankenstein foods" ' (*Daily Telegraph*, 12 February 1999) and 'THE PRIME MONSTER: Fury as Blair says "I eat Frankenstein food" and it's safe' (*Daily Mirror*, 16 February 1999). Concerns have been expressed by many scientists and science policy groups about the potential for public responses to nanotechnology to replicate these earlier experiences (Turner, 2003). FOEA have raised particular anxieties over food-safety issues given that they claim in excess of 100 food, food packaging and agricultural products containing nano-ingredients are currently on sale internationally without mandatory food-labelling measures in place (FOE, 2008a). We argue that journalists' prior experiences of covering such science controversies are likely to colour their treatment of nanotechnology. 'Frankenfood' metaphors are already beginning to be used within the coverage with headlines such as 'Alert over the March of the "Grey Goo" in Nanotechnology Frankenfoods' (*Daily Mail*, 2 January 2008). The Canadian Action Group on Erosion, Technology and Concentration (ETC), which played a key role in turning European public opinion against GM 'Frankenfoods', have switched their major focus from agribiotechnology to nanotechnology. ETC have called for a moratorium on the commercialisation of new nano-scale materials until laboratory protocols and regulatory regimes are in place that take account of their special characteristics and demonstrate them to be safe.

Accordingly, given the degree of interest among scientists and policymakers in the potential of nanotechnologies in various fields (including environmental sustainability, engineering and medicine), a growing number of countries are debating how best to 'engage the public' so as to avoid strong negative public responses, as seen with the 'GM-style' backlash against the technology. As detailed in our chapters, recent science reports and academic discussions on nanotechnologies emphasise the importance of public engagement during the early phase of technological development (i.e. 'upstream'); however, the question of what constitutes such 'engagement' and what the role of the media may be in this process have received only scant attention. In particular, the recent focus on 'upstream' public engagement reveals a linear model of innovation that belies how technologies develop in practice (Joly and Kaufmann, 2008). Despite widespread acknowledgement of the potential significance of the media in influencing attitudes and establishing the agenda for debate, the role of the media in forming public knowledge, we suggest, has been poorly theorised. Scientists often blame 'the media' for 'misrepresenting' nanotechnology and for not adequately conveying 'the science facts'. This, we argue, reveals a simplistic portrayal of science mediation and denies the power relations of science.

Publics' responses to nanotechnology thus far would seem to broadly mirror responses to new genetic technologies in the last decade or so; namely, anticipation of its future benefits (utopianism) mixed with fears of its dangers (dystopianism), especially with unregulated developments. As with developments in the new genetics, which are seen to have the potential to create a new genetic divide, particularly between the developed and the developing world, innovations in nanotechnology are predicted by some commentators to lead to a 'nano-divide' between those who benefit and those who are disadvantaged by technologies (Court et al., 2004). Further, as with genetic technologies, fears about the detrimental impacts of 'tampering with nature' are manifest in depictions of self-replicating 'nano-robots' and 'nano-swarms', as in the case of Michael Crichton's novel *Prey* (2002). Like media portrayals of the risks of GM foods and other genetic technologies (e.g. cloning and embryonic stem cell harvesting), media portrayals of the risks of nanotechnology (e.g. self-replicating 'nano-robots') are seen by some scientists as having the potential to lead to a public backlash against potentially useful applications. Although a public opinion poll by the Royal Society and Royal Academy of Engineering (RS/RAE), published in March 2004, found that 'the overwhelming majority of people had not heard of nanotechnology' ([RS/RAE], 2004), growing news media coverage and other popular media portrayals may generate a context of fear about its implications, and thus constrain developments in the future. In the same way that scientists sought to control responses to cloning in the aftermath of Dolly by using the media to emphasize the benefits of the technology and to draw a distinction between therapeutic and reproductive applications (Petersen, 2002), some scientists have used the media to highlight the potential benefits and safety of nanotechnology, or to downplay the more dystopian scenarios (see Phoenix and Drexler, 2004). Clearly, much may be learnt about the mediation of science, the social production of risk and the public representation of science by analysing public discourses about nanotechnology, its applications, benefits and dangers.

Outline of chapters

This book comprises six chapters, each of which addresses different aspects of the challenge of communicating nanotechnology.

Chapter 2 lays the ground for what follows by examining broad issues about how science journalism is produced in general, including what constitutes 'newsworthiness' as well as credibility where news sources are concerned. It then moves on to address how journalists negotiate competing claims about the possible risks associated with nanotechnologies.

Chapter 3 considers the media politics of nanotech within debates concerning the role of contemporary media in communicating risk. Here we provide a critical survey of the literature within the field and tease out the key conceptual and methodological issues. We examine the processes through which particular themes and voices become dominant or marginalised. In particular, we address the question of why there has been an absence of reference to particular actors and issues, such as economic, ethical, theological and legal issues, and whether a particular framing is contingent on certain events or periods in the policy-making cycle.

Chapter 4 surveys the literature on news media representations of nanotechnology and presents the findings of our ESRC project on nanotechnology and the news. This leads to an examination of why early news media coverage in general has tended to emphasise the revolutionary beneficial applications rather than the potential risks of nanotech, as was the case with earlier biotechnology debates. A range of examples are provided from UK and US studies of nanotech news coverage.

Chapter 5 examines existing work on the public discourses surrounding nanotechnology, supplementing it with new findings drawn from the authors' own research. It begins by examining studies of public perceptions of science journalism (and attendant risk issues), in general, before turning to those devoted to nanotechnology. The chapter presents further data drawn from the authors' recent ESRC study involving interviews with scientists and journalists about their perceptions of public knowledge and awareness and the role of the media.

Chapter 6 explores how scientists and policymakers seek to establish a positive portrayal of nanotechnologies in the face of concerns about developments and an apparent decline in trust in expertise and the effectiveness of regulatory systems. Based upon the findings from a British Academy funded study, it explores both the expectations and the fears that nanotechnologies engender in relation to health and environmental sustainability and how these are communicated.

In the Conclusion, we offer some broader reflections on our analysis of communicating nanotechnology issues, identify gaps in understanding and propose some areas for future work. We point to the need for a more theoretically sophisticated understanding of the science–society relationship that pays cognisance to the socio-political significance of the media and the inescapably mediated character of science.

2
Reporting Science

For the past decade, nanotechnologists have wowed the public with our ability to manipulate matter at the atomic level and with grand visions of how we might use this ability. [However, any] technology that promises so much change is bound to generate controversy, because with such awesome power comes the capacity to push beyond boundaries that society has deemed acceptable. Put another way, societal and ethical concerns can rapidly turn wow into yuck. Kristen Kulinowski.

(2004 : 17)

Science, it is often said, gets a bad press. Explanations for this apparent problem, in the opinion of some journalists, tend to revolve around the charge that most types of science fail the test of newsworthiness. Routine science, they believe, is really rather boring. It lacks the stuff of drama necessary to spark lively newspaper headlines. At the same time, some scientists maintain that on those occasions when a certain scientific development is given due prominence, it all too frequently happens for the wrong reasons. Not surprisingly, they are quick to condemn instances of sensationalist reporting – where news values have given way to entertainment values – for misrepresenting the nature of scientific enquiry, and rightly so.

Science typically appears in the press as 'an arcane and incomprehensible subject', Nelkin (1995) observed in her classic study of science journalism in the US. Surrounding it is a certain 'mystique' that implies it is to be properly regarded as a 'superior culture' with a 'distanced and lofty image'. Far from enhancing public under-standing, she argues, 'such media images create a distance between scientists and the public that, paradoxically, obscures the importance of science and its critical effect on our daily lives' (1995, 15; see also Allan, 2002, 2008; Bell, 2005; Broks, 2006; Bucchi, 1998; Gregory and Miller, 1998). It is precisely this problem where nanotechnol-ogy is concerned that provides us with a rationale for investigating a host of issues in this chapter, ranging from the intense pressures con-straining the reporting process, journalistic perceptions of the relative authority of informational sources and the framing of scientific con-cepts and principles, across a range of media. Even more challenging to discern, however, will be the factors shaping how science is ren-dered newsworthy in the first place – that is, the complex ways in which journalists make their routine, everyday decisions about how to incorporate scientific claims into a story. Journalists themselves will often maintain that they are simply following their 'gut feelings', 'hunches' or 'instincts' when performing this role, a process that in their view is merely a matter of 'common sense'.

This chapter's exploration of science news aims to unravel the largely unspoken, taken-for-granted conventions that underpin these seemingly *ad hoc* judgements. In so doing, it will seek to show that an analysis of the issues involved will enhance our understanding of the challenges currently confronting journalists as they strive to report the complexities at the heart of current debates about nanotechnology.

Science and its publics

Impassioned debates over how best to report on science and technol-ogy in the news media are hardly new, and yet there appears to be a growing sense of urgency on the part of those who seek to speak on behalf of scientific inquiry today. The relationship journalists have with their sources, as will be discussed below, is vital in this regard, for seldom do they find themselves on an equal footing. 'In the case of nanotechnology', Lee and Scheufele's (2006) study of the US news coverage indicates, 'it may be reasonable to assume that the posi-tive nature of coverage so far and the novelty of the technology also help to increase levels of deference toward the authority of scien-tists currently working in this area' (2006, 823). Quite what counts as 'positive' (or 'beneficial') or 'negative' (or 'risky') will depend, of course, on the normative criteria informing competing perceptions. This point is underlined by Pense and Cutcliffe (2007) in their assessment of how different social groups perceive what is at stake. Nanotechnology is a 'disruptive technology', they maintain, 'because there are soldiers, researchers, doctors, patients, sci-fi readers, bikers, and police officers at the table – and, as we might expect, the problems and potentials with nanotechnology are not visible in the same way' to these different stakeholders (2007, 364). Also sitting at this imagined table, of course, are journalists.

Scientists engaged in nanotechnology research are likely to share with their colleagues in other areas of scientific enquiry certain reservations when dealing with journalists. While scientists are usually quick off the mark to identify what they regard as unfortunate misconceptions in public debates about the aims and objectives of scientific research, they are typically reluctant to engage in the task of clearing up these same misconceptions. Increasingly in the eyes of some in the scientific community, however, this reticence is becoming more and more difficult to justify. At issue, in their view, is the fear that support for science among members of the public is in a state of decline, especially among young people. Efforts to support this claim regularly rely on evidence gleaned from opinion polls and surveys attempting to measure the public's understanding of science, the results of which in many Western countries often appear to show a steady increase in antiscience attitudes. Others dispute this interpretation, but in either case much is made of the corresponding decrease in the numbers of students enrolled on science courses at both school and university levels. This apparent trend is regarded by some commentators as having alarming implications for the next generation of science teachers and researchers, as well as for the future quality of modern society's industrial and technological development more generally.

Attributions of blame for this apparent disillusionment, if not outright antipathy, with science are being laid at the doorstep of a number of different possible sources by pro-science advocates. Typically singled out for criticism by pro-science groups are voices from the arts, humanities and social sciences, many of which are regarded as exemplifying the worst sorts of tensions illuminated so sharply in C.P. Snow's (1965) *The Two Cultures* of several decades ago. Particularly worthy of censure today, in the opinion of these critics, are those social and cultural theorists who subscribe to the philosophical tenets of 'postmodernism' as a distinct intellectual position (see also Best and Kellner, 2001). Postmodernists, they argue, are deeply misguided in their questioning of the universality of reason, especially when they challenge the precept that there exists an external reality to be rationally detected through scientific modes of investigation. For these critics, the postmodernist dismissal of science as merely one way to understand the world, a 'language game' intrinsically no better or worse than others, is nothing short of contemptuous. Hence their outrage at what they perceive to be the growing influence of postmodernist attacks on science, not only in the educational system but increasingly in the media. The failure of postmodernists to engage with science in an adequate manner, these critics insist, threatens to turn public scepticism about traditional standards of truth into bleak, nihilistic forms of alienation.

Pro-science groups also express concern about the various forms of 'pseudo-science' regularly being presented in the media. They are especially critical of those forms of popular culture which proclaim for themselves (explicitly or, more typically, implicitly) a scientific status that is completely unwarranted. For example, the 'meaningless pap' of astrology is offensive, in Dawkins' (1998) view, 'especially in the face of the real universe as revealed by astronomy', and also because of the 'facile and potentially damaging way in which astrologers divide humans into 12 categories' or 'signs' under the horoscope's star system (1998, 115, 118). Considering one's horoscope in a newspaper, for some readers a light-hearted form of amusement, can have unfortunate consequences for the vulnerable, who are anxious to follow the advice on offer. Similarly, he cites the media's 'obsession' with the paranormal, taking strong exception with 'second-rate conjurors masquerading as psychics and clairvoyants' (1998, 115). From there, he contends, it is arguably a short step to believing in telepathy and magic, or even entertaining the possibility that spirits and hobgoblins exist. 'Disturbed people recount their fantasies of ghosts and poltergeists', he writes, '[b]ut instead of sending them off to a good psychiatrist, television producers eagerly sign them up and then hire actors to perform dramatic reconstructions of their delusions – with predictable effects on the credulity of large audiences' (1998, 129), and, it may be argued, with a corresponding impact on what comes to be understood as 'science' as a result.

Moreover, it is often the case that these sorts of criticisms revolve around the perception that science is being effectively 'dumbed down' in the name of making it more publicly accessible. Dawkins sees in the 'science literacy' movement in countries like Britain and the US, for example, an anxiety among scientists to be loved by members of the public. As he writes,

> Funny hats and larky voices proclaim that science is fun, fun, fun. Whacky 'personalities' perform explosions and funky tricks. I recently attended a briefing session where scientists were urged to put on events in shopping malls designed to lure people into the joys of science. The speaker advised us to do nothing that might conceivably be seen as a turn-off. Always make your science 'relevant' to ordinary people's lives, to what goes on in their own kitchen or bathroom. Where possible, choose experimental materials that your audience can eat at the end. [...] The very word science is best avoided, we were told, because 'ordinary people' see it as threatening.
>
> (Dawkins, 1998, 22)

Calculated efforts to achieve this 'dumbing down' such as these ones may indeed attract the interest of the public, he argues, but they can all too easily become condescending and patronising or, even worse, turn into a kind of 'populist whoring that defiles the wonder of science'. His choice of the latter phrase, in our view an inappropriate one, is in any case indicative of the passion with which he declares that 'we shouldn't need whacky personalities and fun explosions to persuade us of the value of a life spent finding out why we have life in the first place' (1998, 23). Thus while disputes about the 'dumbing down' of science have been around for a long time, Dawkins's voice is part of a growing chorus from within the scientific community demanding a counter-attack of sorts against what they perceive to be an increasingly dangerous misrepresentation of science in the public sphere.

In relation to nanotechnology, several studies have pointed out that the prevailing themes in the relevant news coverage tend to accentuate the perceived benefits to be gained over and above the possible risks (Lee and Scheufele, 2006; see also Anderson et al., 2005; Cobb, 2005). The extent to which this is the case today remains a matter of debate, however, with some research suggesting that news reporting of its social implications has tended to be negative, especially in its early phases (see Friedman and Egolf, 2005; Stephens, 2005; and Chapter 4 of this volume). Scientists Wood et al. (2008) observe, 'may be slightly bemused by the way in which nanotechnology has reached the popular press – at least before it truly crystallized – as a potentially dangerous development, alongside the scares about genetically modified organisms (GM) and past disasters such as asbestos, thalidomide and BSE' (2008, 13–14). Even where the term 'nanotechnology' itself is concerned, some studies suggest that it is hardly 'neutral' in its everyday usage. Randles et al. (2008) contend, for example, that 'there is much evidence to support the argument that its widening into popular parlance is part of a socio-technical and above all political project to mobilize actors around the objective of raising finance'. The effect of this mobilisation, they believe, is to channel 'substantial levels of investment capital and attention towards particular privileged projects' (2008, 3).

Indeed, as we will argue in the later chapters of this book, this imperative is particularly significant where issues of 'risk' are concerned in news reports of nanotechnology. A related study of US newspaper reporting by Stephens (2005), for example, suggests that news articles that 'emphasize the benefits versus the risks of nano are more likely to be found on the business pages or in the special science and technology sections of newspapers'. At the same time, he adds, articles emphasising 'the risks of nano are more likely to appear on the front page or main news section, on general features pages, or in sections where movies and books are reviewed' (2005, 197; see also Throne-Holst and Stø, 2008). Indeed, it is the perception of possible risks associated with nanotechnology – that is to say, the likelihood of harm occurring – that helps to explain journalists' interest in nanotechnology in the first place. At the same time, however, this benefits versus risks calculation is fraught with difficulties. 'Engineered nanomaterials are a new class of materials', Bell (2006) observes, 'and definitions are not standardized, mechanisms are unfamiliar and exotic, and unknowns abound' (2006, 1). It is hardly surprising, then, that journalists confront particular challenges when seeking to determine what is newsworthy about nanotechnology.

Making science newsworthy

'Since the earliest days of science writing', Friedman (1986, 17–18) observes, 'the profession has been beholden to the two worlds of science and journalism, functioning under the rules and constraints of both'. Moreover, she adds, 'the influence of these two worlds has not been equal', with those involved from the journalistic side playing a more decisive role in shaping the development of science writing to meet their respective needs. This contest continues today, of course, as the status of science writing as a distinct strand of journalism is being increasingly recognised across a range of media genres.

'Over the last 30 years or so', as science reporters Blum and Knudson (1997a, ix–x) point out in a US context, 'science writing has been transforming itself into something beyond a strange little subculture of journalism', namely as a profession in its own right (see also Friedman et al., 1986). Driving science writing as a profession, in their view, is first and foremost science itself:

> the post-World War II boom in research, the space race of the 1960s, the technologies of today that are opening the subatomic and molecular worlds at a still-dizzying pace, giving rise to a revolution in personal communications and in our knowledge of genetics and biology. Such discoveries have altered the world we all live in, and it has increasingly fallen to the media to explain the new technologies and report on their impact, good and bad.
>
> (Blum and Knudson, 1997a, x)

Where science journalists once acted primarily as 'scouts' on a reconnaissance mission, 'trying to bridge a big divide by bringing back messages from one side to the other', today they are unlikely to be content to limit their work to this role alone. Although they still need to perform the task of translating the abstract complexities of scientific enquiry for their audience, Blum and Knudson (1997a, x) maintain that the science reporter's role must also stretch to encompass a larger sense of public responsibility. In their words:

You can paint an awesome and adventuresome picture of space exploration with all its glittering planetary rings, but you can also acknowledge its risks and probe its failures. You can point out the medical and agricultural benefits of the new biotechnology or the mapping of the human and other genomes; but you can also question what harm may come of the new knowledge and capabilities, discuss what safeguards will be put in place, and talk about how much big science costs and who pays for it.

(Blum and Knudson, 1997a, x)

It is this shift from 'translating' science for members of the public to making it relevant to them and their daily lives that poses a particular challenge. Science writing, for many of its practitioners, is one of the most deeply satisfying forms of journalism to learn, and yet is also one of the most difficult to get right.

Indeed, of the various beats newspaper reporters regularly cover, the science beat is one of the most challenging. 'In most other speciality beats', as Rensberger (1997, 8) of the *Washington Post* points out, 'reporters become familiar with a modest body of knowledge (how a city council functions, for example, or the rules of baseball) and turn to the same few, first-name-basis sources every day'. The science reporter, he adds, seldom enjoys this luxury, having instead to come quickly up-to-speed on a host of emerging events or issues as they surface from one day to the next. Breaking science news is very difficult to anticipate: 'Today the story may be a claimed advance in treating cancer, tomorrow it may be explaining atmospheric chemistry and, for the weekend, the latest experiment in fusion power research' (1997, 8). As quickly as the issue changes, of course, so will the reporter's perception of which potential sources of information are likely to be most appropriate for the news item being prepared. In the case of nanotechnology, as will be shown in Chapter 4, a wide array of beats comes into play beyond those revolving around science. These include technology, environment and health, amongst others.

One of the telling realities of writing about science for a daily newspaper, according to Rensberger, can be found in the mailroom. There, he argues, one will typically find that the 'mailboxes of the science and medical reporters will be among the most stuffed' (1997, 7). Demands on the science reporter's time and attention are as constant as they are vociferous. In Rensberger's experience, such demands are likely to arise from individuals and groups in all walks of life, but especially from people associated with 'universities, corporations, think tanks, government agencies, advocacy groups, independent research institutions, museums, public relations agencies, hospitals, and scientific journals' (1997, 7). The sheer volume of these efforts to capture the newspaper science reporter's interest necessarily means that he or she has to assume a 'gatekeeper' function in the newsroom. That is to say, the science reporter will have to decide on a routine basis 'what developments in the real world get into the news, and hence reach the public'. This gatekeeping function is contentious, not least where the reporter's proclaimed objectivity is concerned. 'After all', Rensberger (1997, 11) argues, 'of the dozens of stories we could do on any given day, we reject most or all possibilities. In this we exercise our opinion as to what is a good story' (see also Kiernan, 2006; Malone et al., 2000).

And what makes a good science story, in Rensberger's opinion? In essence, he maintains that the following five criteria – in combination and to varying degrees from one story to another – are especially valued by newspaper science journalists:

- *Fascination value*: Rensberger (1997, 11) writes, 'This is the special commodity that science stories, more than any other kind, have to offer. People love to be fascinated, to learn something and think, "That's amazing, I didn't know that".' By this criterion, then, dinosaurs 'may be the quintessential fascinating topic for science writers'.
- *Size of the natural audience*: here Rensberger is referring to the number of newspaper readers who are already aware that they are interested in following a news story about a given topic. 'If the story is about a common disease that everyone had had or fears getting', he declares, 'the natural audience will be larger than for a rare disease' (1997, 11).
- *Importance*: the subjective quality of any attempt to assess importance is readily acknowledged by Rensberger, although he suggests that to 'judge a story idea on this point, you would try to decide whether the event, or finding, or wider knowledge of the event or finding is going to make much of a difference in the real world, especially in that of the average newspaper reader' (1997, 11–12). Following this logic, 'AIDS is important, bunions are not.'

- *Reliability of the results*: to pinpoint this criterion, Rensberger poses the question: 'Is it good science?' The single most useful guideline for determining reliability, he argues, is science's own peer review system. This system, he writes, 'is a time-tested way to minimize the odds that a misunderstanding is promulgated to the world at large. Science writers who ignore the system risk misleading their readers and embarrassing themselves' (1997, 12; see also Young, 1997).
- *Timeliness*: 'The newer the news', Rensberger (1997, 13) states, 'the newsier it is'.

Each of these criteria directly inform news judgements about nanotechnology, but questions regarding the 'reliability of the results' are likely to take on an added significance. Responsible coverage requires accurate understanding, as Bell (2006) observes, which can be especially challenging in such a rapidly developing area characterised by so much uncertainty. Here she offers an illustrative example:

> In 2005, popular articles reported on a study that asserted that alumina (aluminum oxide) nanoparticles in soils appeared to slow the growth of plants – possibly important for environmental disposal. What the scientific report failed to state, however, is that alumina dissolved in solution is highly toxic to plants. So the observed toxicity may have been irrelevant to engineered NSPs [nano-scale particles]. In other words, even though journalists had accurately reported the paper's findings, the scientific paper itself was faulty in ascribing cause and effect – and those deficiencies were magnified in the popular press.
>
> (Bell, 2006, 6)

Bell stresses how important it is for journalists to contact scientists to discuss their findings, to really push and prod the conclusions being made. Journalists need to ask, for example: 'Is this substance also toxic in different forms or in solution? Are the effect(s) you report unique to its nanostructure? What do skeptics say about these conclusions?' (2006, 6). At the same time, journalists need to gather the views of other researchers so as to help determine the reliability of a given study's findings. Moreover, this process of checking and double-checking results will similarly help to determine the relative news value of a study, in part by ascertaining perceptions of its significance (or otherwise) within the community of nanotechnology researchers.

News values

News coverage of science tends to favour certain areas of scientific enquiry over and above other areas, a pattern which is usually observable from one newspaper or news broadcast to the next. The process by which certain scientific developments are rendered newsworthy while others, in contrast, are deemed unworthy of attention is the outcome of a complex array of institutional imperatives. Journalists, together with the other individuals involved in the work of processing news in a particular news organisation (editors play a key role here), bring to the task of making sense of the social world a series of 'news values'. These news values are operationalised by each newsworker, as Hall (1981) suggests, in relation to their 'stock of knowledge' about what constitutes 'news'. If all 'true journalists', he argues, are supposed to know instinctively what news values are, few are capable of defining them:

> Journalists speak of 'the news' as if events select themselves. Further, they speak as if which is the 'most significant' news story, and which 'news angles' are most salient, are divinely inspired. Yet of the millions of events which occur every day in the world, only a tiny proportion ever become visible as 'potential news stories': and of this proportion, only a small fraction are actually produced as the day's news in the news media.
>
> (Hall, 1981, 234)

Hence the need to problematise, in conceptual terms, the operational practices in and through which news values help the newsworker to justify the selection of certain types of events as 'newsworthy' at the expense of alternative ones. To ascertain how this process is achieved, researchers have attempted to explicate the means by which certain news values are embedded in the very procedures used by reporters to impose some kind of order or coherence on to the social world (see also Allan, 2004).

The news values associated with sensationalised 'gee-whiz' or 'weird-and-wacky' moments differ markedly from those indicative of the more routine types of science stories likely to appear in the daily news cycle. To better understand these distinctions, researchers have attempted to explicate the means by which certain news values are embedded in the everyday procedures used by science reporters to justify the selection of certain types of events as newsworthy at the expense of alternative ones (see also Allan, 2002; Bucchi, 1998; Friedman et al., 1999). Hansen (1994) suggests, on the basis of his study of science reporting in the British press, that the 'most pronounced criterion of newsworthiness is whether science can be made recognizable to the reader in terms of human interest or in terms of something readers can relate to' (1994, 114–115). Particularly prized, as a result, are those events that illuminate the relevance of science to daily life, enabling the journalist to adopt a 'human angle' when constructing the news story. Further factors informing this process of negotiation include the efforts made by news sources or stakeholders themselves to influence journalistic judgements, as well as the relative complexity of the event itself. 'The more complex or inaccessible a piece of science news is', Hansen writes, 'the more "translatory work" it will require on the part of the journalist to make it intelligible and interesting to the readers'. Time is of the essence for journalists working to conform to a daily production schedule, especially where deadlines are concerned. That said, while a significant scientific 'breakthrough' may be judged to constitute 'hard' news, and thereby warrant immediate coverage, it is much more likely that the science involved will be, 'in news terms, a slow process of small incremental developments' (1994, 115).

The norms of science journalism, it follows, seldom align comfortably with those of the science being covered. Wilcox's (2003) research identifies what she describes as the 'hype/space dilemma' shaping this alignment. Specifically, she highlights some of the contradictions encountered by journalists when scientific claims need to be hyped in order to secure the necessary space in the newspaper. '[T]he media require scientific studies to provide dramatic new findings and dramatic conflicts', she writes, 'while the conventions of scientific journalism require that the results of single studies be de-emphasized in favour of the scientific context of the research' (2003, 243). Lynch and Condit (2006), in their assessment of this dilemma, suggest that it revolves around a basic conflict, namely the journalist's struggle to make a story interesting, even sensational, in relation to other stories competing for the same space while, at the same time, reaffirming professional ideals of truth-telling and balance. Precisely how this conflict is negotiated will have significant consequences not only for what gets covered, but also for the standards of accuracy applied when scientific terminology is being translated into layperson vocabulary. 'At best', Condit (2004) points out, 'this standard includes fidelity to sources, a balance among and inclusion of different viewpoints, and a translation that conveys some main idea from a study clearly'. What more dispassionate, thorough reporting loses with respect to excitement, it gains by helping to ensure stories avoid 'the kind of distortions that might encourage inappropriate (or even dangerous) behaviour or unrealistic expectations' (2004, 1415).

News values are thus context-specific, and as such never fixed once and for all – instead they are always evolving over time, being inflected differently from one news organisation to the next. Still, the ostensibly 'common sense' criteria informing definitions of newsworthiness in science journalism have proven to be surprisingly consistent over the years. 'For all of modern science's sophisticated concepts and technology', Young (1997, 114) observes, 'journalism's traditional five *W*s and an *H* – who, what, when, where, why and how – still form the core of science reporting'. Toner (1997) agrees, but with an important caveat:

> Editors insist – and many believe – that busy readers have no time for the 'rest of the story.' The five *W*s that journalists once revered are often reduced to four. Yes, we can fit the who, what, where and when into the little space on the front page. But skip the why. That fifth *W*, after all, may only raise more questions than it answers.
>
> (Toner, 1997, 130)

This is the key issue, in Toner's view. Science journalists who fail to ask the question 'why' are too often providing information that satiates people's curiosity, as opposed to stimulating it. 'Like a weed', he writes, 'curiosity has a habit of popping up in the wrong place. It can be unruly and hard to control. It is robust and tenacious. And where one question sprouts, many more are bound to follow' (1997, 128). It is this contagious aspect of curiosity, in his opinion, that renders it 'such a powerful tool for journalism'. Moreover, he contends, 'Readers, listeners, and viewers may appreciate our wit, our incisive grasp of complex issues, and the clarity of our delivery, but by planting the seeds of curiosity, we make our audience accomplices in the pursuit of knowledge' (1997, 129).

If science items can often be the 'furthest thing from breaking news,' as Petit (1997) suggests, 'this can be their charm' (1997, 187). Here he draws a sharp contrast between the kinds of news stories he typically writes and the more usual kinds of items published alongside them in the same newspaper. 'Like stories in astronomy, or on fossils of prehistoric people', he observes, 'discoveries about the Earth's history and behavior provide for many people a welcome and invigorating break, a mental escape from the daily diet of human disaster, political skulduggery, and crime news' (1997, 187). Ropeik (1997), a television news reporter covering science and environmental stories, writes,

> The feedback I've gotten in my 18 years as a journalist leads me to believe that news consumers are curious. They *want* to know. They *want* complicated things explained to them. They have a gee-whiz button waiting to be pushed. I look for ways to push it, in how I organize, how I write, and literally in the tone of voice I use as I narrate my stories'.
>
> (Ropeik, 1997, 38)

As a result, Ropeik places a considerable emphasis in his reporting on being 'simple and clear, given the brevity and mind-numbing nature of television news' (1997, 38). To be effective means playing to television's principal strength, that is, 'to let the pictures do the talking'. The rendering of complex scientific issues into visual images, when done thoughtfully, can spark and sustain the viewer's interest in a way that helps to make core points register. The 'power of pictures to tell the story', he maintains, 'is the television newsperson's greatest tool' (1997, 38).

This issue evidently strikes a resonance with Flatow's (1997) experience in television reporting. In his discussion of the production of long-format science stories for television, he observes,

> The most challenging stories to produce are the ones in the field of science that people 'think' they don't find interesting like physics or chemistry. Just the mention of the words can send viewers to

their remote controls. The trick, then, is to disguise the science in the piece, hide it, and spring it on the viewer suddenly. Do this by treating the topic not as a 'science' story but as an Agatha Christie mystery. Your scientists are not white-coated laboratory technicians but science sleuths on the trail of a suspect. Nobody can resist a good whodunit.

(Flatow, 1997, 40–41)

The search for such dramatic elements, he concedes, is difficult. 'It may also mean', he adds, 'finding scientists who are good story-tellers, personable on camera, and willing to submit to the rigors of television' (1997, 41). This last point raises a host of issues for the unsuspecting scientist, as Flatow proceeds to elaborate:

scientists who agree to become television 'talent' may have no idea of the demands that may be made of their laboratory. Dozens of phone calls interrupt their work. Scripts have to be written and re-written. Schedules must be arranged. Then comes the invasion. Laboratories are besieged by hoards of camera, lighting and sound people. All work stops while those 'TV people' take over. Unsuspecting scientists may balk at the commotion and decide that this is not what they bargained for.

(Flatow, 1997, 41)

The logistics involved in finding scientists willing and able to serve as news sources can be formidable. When asked to reflect on how they go about their daily work of identifying those 'newsworthy' sources deserving to be included in a news account, journalists will often claim that they simply follow their 'gut feelings', 'hunches' or 'instincts'. Many insist that they have a 'nose for news', that they can intuitively tell which sources are going to prove significant and which ones are bound to be irrelevant to the news item. It is this issue of how science journalists interact with scientists, then, that is the focus of the next section.

Scientists as sources

The phrase 'Get all sides of the story' signifies a vital tenet of journalism that Greenberg (1997), a newspaper editor, emphasises in his discussion of science reporting. 'Whether a science story involves long-term investigation or quick-turnaround breaking news', he writes, 'it requires well-rounded, balanced reporting that relies on the savvy and expertise of the writer perhaps more than most other beats' (1997, 97). He proceeds to identify several basic sources for reporting science news stories as follows:

- *Journals*: the steadiest of science news sources, in Greenberg's view, journals encourage a certain form of 'pack reporting' among journalists. The reason, he argues, is straightforward enough: 'If a researcher is going to drop a bombshell, chances are it will land smack in the pages of *Nature*, *Science*, *JAMA* the *New England Journal of Medicine*, *The Lancet*, or one of scores of other smaller but respected publications' (1997, 97). Publications such as these ones are routinely monitored by the major news organisations, namely because to miss an important research story is 'an acknowledged sin' (see also Young, 1997).
- *Meetings*: after journals, scientific meetings are the second likeliest place a scientist will announce a major research finding. 'This does not happen as often as it used to', Greenberg (1997, 97) argues, 'primarily because refereed journals are considered by many a purer forum in which to divulge advances'. Such meetings are thus less likely to generate 'breaking news', but are nevertheless worth attending by science journalists for 'background' and to 'cultivate sources' for interviews.
- *Breaking news*: just like other journalists, Greenberg points out, science reporters work 'at the whim of events'. Speaking of the situation in Los Angeles, where he is based, he states, 'we're poised to cover an earthquake every day, and to a lesser extent fire and floods. For science writers it is especially important to develop sources expert in such local phenomena that are likely to recur' (1997, 97).
- *Press conferences* or *press releases*: information gathered from these sources needs to be treated with caution. Greenberg argues that a wary eye should be trained on any institution or scientist publicising their research prior to journal publication, not least because there are 'many real-world agendas that sneak into science, like funding, competition, ambition, and glory' (1997, 97). That said, though, he does concede that occasionally press

announcements, or contacts from public relations, yield what he considers to be a legitimate news story, even an exclusive one. As he recalls, 'This was the case recently when, through a long, close working relationship with a public relations person at a local hospital, the *Los Angeles Times* was present and broke the story of the first gene therapy procedure on a new-born' (1997, 97).

• *Unsolicited calls*: Greenberg expresses his advice where this type of source is concerned in blunt terms: 'If "scientists" you don't know want to publish their original "research" in your newspaper or magazine, rather than in a journal, run for cover' (1997, 97–98).

• *Own sources*: a crucial task for any science reporter, indeed for Greenberg possibly the most important part of covering science, is 'building up a cadre of reliable, informed sources that you can call for reaction and comment – both on or off the record – about any range of stories' (1997, 98). Such a task takes time and experience, he argues, as well as 'gaining the trust of such people through consistently accurate and well-written stories' (1997, 98).

Overall, in Greenberg's view, the value of a science news item will ultimately come to rest on the calibre of the sources the reporter has drawn upon to tell the story. In almost every instance, he maintains, the nature of the story determines where the reporter looks for appropriate sources. Such is not always the case, though. 'While the rules for choosing sources on most science stories are clear', Greenberg (1997, 99–100) writes, 'certain kinds of stories can render those rules as gooey as volcanic mudflows.' Here he cites 'the infamous "cold fusion" experiment' of 1989 to illustrate his point, an occasion where despite the fact that many journalists were deeply sceptical of the claims being made they nevertheless gave the announcement blanket coverage as a major news event (see also Bucchi, 1998; Toumey, 1996).

The challenge of balancing as many authoritative sources as possible within a news account appears to be a crucial one for the science reporter. Disagreement between sources cannot always be framed in terms of 'right and wrong', instead sometimes it may be more accurately characterised as simply representing varied interpretations.

'Many scientific studies', Young (1997, 116) maintains, 'are so complex, so difficult to do, that their findings do, indeed, lend themselves to two or more interpretations'. The opinion of the research team, he adds, should not be enshrined in truth when others working in the same field may – quite rightly – view the results rather differently. This point is similarly addressed by Harris (1997), an environmental reporter, who argues that in the case of 'extreme' points of view among potential sources, even greater care needs to be taken to maintain journalistic balance. In his view, the kinds of voices heard at the margins of a scientific consensus tend to be either ignored entirely or treated with equal weight in a news story. What is much more appropriate, he maintains, is for the journalist to scrupulously situate each voice within the larger spectrum of opinion so as to enable the news audience to understand their arguments in context. It is important to bear in mind, Harris (1997, 170) points out, that the 'minority view isn't necessarily wrong – just ask Galileo'.

What motivates the scientist to take the journalist seriously? In Rensberger's (1997, 9) experience, they simply 'want the public to know, to understand, and to be on their side in a world too often given to ignorance, fear and superstition'. In discussing how science reporters interact with their sources, Salisbury (1997) argues that the gap between them is usually not as wide as it tends to be in other areas of journalism. Indeed, in his experience, members of the scientific and journalistic communities are in many ways 'natural allies', sharing as they do 'a skeptical approach to information and a devotion to discovering the truth' (1997, 222). Both sides are likely to benefit from what can be a mutually advantageous relationship. Just as news organisations often seek to boost their audience figures by drawing on reports of exciting scientific discoveries, so scientists can attract political and economic support for their research by receiving favourable media treatment. That said, however, this symbiotic relationship can quickly become problematic at a number of interrelated levels. Indeed, as Salisbury elaborates,

> Scientists live in a different 'time zone' from reporters. They work on projects for months and years, so a paper that has been out for six months can still seem new to them. Conversely, in some fields it takes six months to a year for a paper to appear in a journal, and by then the scientist has moved onto another topic and considers

the paper old news. Because scientists' time frame is so different, they are unlikely to contact the news office at a journalistically appropriate time.

(Salisbury, 1997, 220)

In attempting to overview some of the more pronounced ones, Salisbury observes,

- To scientists, the devil is definitely in the details, while journalists are interested primarily in the big picture.
- To scientists, disputation is part of the process of advancing understanding ever closer to truth; to journalists, conflict is the source of drama that adds zest to a story.
- Scientists are continually trying to build consensus, while journalists focus on the drama of pro and con.
- To scientists, peer review is an integral part of a process designed to reduce errors. To most journalists, allowing sources to review material before publication is an unacceptable ceding of editorial independence.
- To scientists, technical terms provide added precision and clarity to discourse. To journalists, technical terms constitute a jargon that obfuscates science and makes it incomprehensible to the general reader.

(Salisbury, 1997, 222)

Given the significant number of scientists who feel that they have been 'burned' in some way by the media, Salisbury observes that it can be difficult to persuade them that 'the overall benefits to the science community at large are worth the time, energy, and risk involved in dealing with reporters' (1997, 223).

To be effective, then, science journalists need to cultivate a relationship of trust with their sources, a point which we explore in Chapter 4 where nanotechnology is concerned. Here, though, it is important to emphasise how quickly the symbiotic nature of the scientist–journalist relationship can come unravelled under certain circumstances, especially where controversial claims are involved. Speaking from his experience as a science reporter, Young (1997) warns of the 'hidden agenda' a scientist can bring to bear in an interview with a journalist:

The possibility of profiting from their research or a fear of losing their research funding may skew their comments or color their judgement of their work's potential for benefiting humankind. Or they may be wedded too intensely to some cause, such as saving the environment or preventing child abuse, or even to some scientific theory.

(Young, 1997, 115)

This cautionary note similarly informs Trafford's (1997) discussion of how health reporters handle their sources. She proceeds to highlight the danger of quoting 'everyone you talk to on an equal basis in the name of "balance" [when] what you're really striving for is fairness and accuracy' (1997, 137). As she points out, not every source is equal, nor should it be the case that different viewpoints be reported in equal terms: 'The reader expects you to make the first cut in evaluating the major points in a story' (1997, 137). When it comes to handling potential news sources, then, care needs to be taken to evaluate their relative merits. Public health stories, in her experience as a health editor on a daily newspaper in the US, typically start with government officials:

In many ways, they [government officials] are like real estate agents: they are often friendly, knowledgeable, and sophisticated. They show you a lot of properties. They want you to be happy and they answer a lot of your questions. But remember, they are always working for the seller – namely, the government, and in some cases, the president, or the governor or major who gave them their job.

(Trafford, 1997, 136)

Trafford (1997) discerns several rings of sources flowing outward from this governmental realm of public health officials. The first such ring is composed of major institutions, such as schools of public health, hospital systems, medical schools and research and policy centres. The second ring encompasses advocacy groups, including disease organisations and grassroots citizens' groups, as well as lobby groups, which include promotional foundations and trade associations (such as those associated with hospitals or drug companies). In the third ring are what Trafford calls 'the bystanders', that is, 'those individuals who are more affected by a public health problem and the government's plan to deal with it' (1997, 136). Finally, in the fourth and outermost ring is the general public.

For those science journalists committed to ensuring that their reporting fulfils this sense of public responsibility, a range of exigent issues emerge. Particularly relevant here is the danger that they will be accused of allowing the 'subjective' opinions of their sources to cloud what should be 'objective' statements of fact. But how can the science reporter establish which facts are 'objective' when their sources disagree? Describing the 'hardships and pitfalls' in science writing, Perlman (1997) observes,

> Journals can be filled with deadly jargon. Claims for statistical significance from randomized double-blind clinical trials can be difficult to challenge. Explaining quarks, subatomic particles with their arcane attributes of strangeness and spin and color and charm, may not gladden the hearts of editors or command the column inches that every science writer knows such stories truly demand. And controversy can arise in every field of science to challenge a reporter's confidence in his or her judgement.
>
> (Perlman, 1997, 4–5)

A science editor for a daily newspaper in the US, Perlman is adamant that the science journalist is 'not entitled to bias or conflicts of interest' in their reporting. In the course of making their judgements about what counts as a 'fact', they must eschew hype and be sure to uphold rules of fairness in handling their sources. That said, however, Perlman (1997, 5) insists that 'they must always recognize that merely counting yeahs and nayes in a scientific controversy fails to serve the public and is rarely a guarantee of fairness'. In his view, good reporting entails more than writing about scientific discoveries or developments in a balanced manner. It also means explaining 'their potential impact and their costs and benefits, even while we present the valid sides of the controversies they generate' (1997, 6).

Blum and Knudson (1997b), both science reporters, point to 'the continuing culture conflict between scientists and journalists', arguing that it has 'intensified as popular reporting on science has become more extensive and more influential' (1997b, 76). The relationship between both groups is slowly evolving, however, as the gap between their respective perceptions arguably narrows. Still, according to Blum and Knudson:

> Many [scientists] remain wary of the media: they don't want to look like show-offs; they aren't certain the audience will understand; and when the reporter ambles in and asks whether CO_2 floats or swims, they aren't sure the reporter gets it either. The whole experience can simply make a researcher nervous, and the result is sometimes obsessive focus on perfecting every detail.
>
> (Blum and Knudson, 1997b, 76)

This is not to deny, though, that some improvements are being made. Blum and Knudson suggest that many scientists are beginning to gain a better understanding of the journalistic profession, and in so doing 'may gain a more realistic expectation of media coverage'. Scientists and journalists, they add, share at least one common goal: 'to make science vivid, real, compelling, and important' (1997b, 76).

This is anything but simple in practice, of course. Jarmul (1997, 124), a science journalist, points out how reporters are trained 'to be objective and to keep yourself out of the story' while, at the same time, science teaches them to 'write in an impersonal, passive voice'. Hence his call for science reporters to forget these rules, or at least to re-examine them, in order to become a 'real person' who can 'connect with readers' hearts, not just with their heads' (1997, 124). This need to form points of connection is particularly important where the framing of scientific controversy is concerned, the subject of the next section's discussion.

Framing controversy

A growing number of commentators are expressing their alarm about how economic pressures on news organisations are forcing some to scale back their commitment to science reporting, cutting jobs and refocusing priorities around non-specialist topics. This decline is particularly apparent where investigative reporting of science is concerned (see Allan, 2002; Anderson et al., 2005; Bucchi, 1998; Gregory and Miller, 1998; Hargreaves et al., 2003; Hotz, 2002; Kiernan, 2006; Priest, 2005; Reed, 2001). Dean (2002), former science editor of *The New York Times*, identifies three further difficulties. First, she notes, science journalism's reach has to be remarkably broad. 'We cover everything from anthropology to astrophysics to atherosclerosis. We advise other departments when a ballplayer is injured or a court overturns a pollution regulation.' Second, given that science is becoming increasingly specialised, 'it is hard for journalists, even journalists with advanced training, to know what is important and what is not important'. Third, she points out, scientific research is becoming increasingly commercialised, creating widespread conflicts of interest. As more and more researchers are turning 'their labs into test beds for their own companies, or have grants from major commercial concerns, or seek venture capital', she writes, 'they have powerful motives for making the most of their results and playing down anything that might challenge them' (2002, 25).

The relationship between journalists and scientists, some suggest, is becoming increasingly fraught. 'More than ever', Hotz (2002) contends, 'scientists aggressively court media attention, even as – paradoxically – unprecedented commercial secrecy comes to shroud so much of what scientists do today and financial conflicts of interest among researchers have become so common' (2002, 6). Scientific publications, including several of the leading peer-reviewed journals, have become caught up in this commercialisation, which leaves journalists wondering which sources they can trust. This problem can be further compounded at times by what is described as the embargo system, whereby journals send out advance details about scientific research to journalists on condition that they withhold reporting on them until the embargoed time elapses. Kiernan's (2006) analysis of its impact demonstrates how these journals succeed in exercising considerable control over the range and diversity of science stories, with their editorial judgements having an especially strong influence on non-specialist journalists struggling to cope with scientific complexity. Nonetheless, it is important to recognise that this occurs with the full co-operation of those journalists who prefer a 'level playing field' with rival news organisations. Consequently, he writes, 'if journalists are in a stranglehold, it is a self-inflicted stranglehold – and one that does not serve the public interest' (2006, 122). Whitehouse (2007) concurs, contending that this 'cosy and secretive arrangement' acts as a 'marketing tool' for the journals while it simultaneously 'encourages lazy reporting and props up poor correspondents'. Genuine scoops, as a result, 'are few and far between in science as the embargo system militates against them'.

Several researchers investigating the ways in which science journalists put together news stories have underscored the importance of examining the actual language used to make complex scientific findings meaningful in the 'lay press'. Such a process will necessarily entail moving beyond direct dependence on scientific discourse so as to find more suitable modes of expression, sometimes requiring the journalist – as Friedhoff (2002) maintains – 'to think about communicating ideas and concepts that in many cases were previously unthinkable'. Findings, she adds, 'need to be embedded into a commonly understood context' (2002, 22–23). In order to unravel this process of embedding, some researchers have utilised the concept of 'framing' to help specify how science stories tend to respect certain 'principles of organisation' (Goffman, 1974). That is to say, framing is often described as a strategy employed by journalists to define the nature of an occurrence as a meaningful event. In the words of Nisbet and Mooney (2007):

> Frames organize central ideas, defining a controversy to resonate with core values and assumptions. Frames pare down complex issues by giving some aspects greater emphasis. They allow citizens to rapidly identify why an issue matters, who might be responsible, and what should be done.
>
> (Nisbet and Mooney, 2007, 56)

Framing, it follows, is the subject of intense negotiation, and as such frequently proves to be a contested process – not least between journalists and their sources (as well as between journalists and their editors within a news organisation, where decisions about placement, headlines, images, captions and the like become important). Indeed, in addition to helping to define what is significant to know, frames have far-reaching implications for how claims made by sources are selected (or not) as newsworthy, the narrative conventions guiding the ways in which they are reported, and the possible consequences for influencing public perceptions. Sources themselves are often acutely aware of the importance of the framing process, so will make every effort to try and ensure that their preferred definition of the issue or event is placed in a positive light.

To clarify the nature of the tensions which can arise in this framing process, it is worth considering it from the vantage points of the science journalist, on the one hand, and the scientist as a prospective source, on the other. Of the myriad of concerns confronting both of them respectively, several are particularly pertinent. Turning first to the journalist, he or she will likely want to determine the answers to questions such as the following:

- Who is a credible, trustworthy and legitimate scientific source for me to contact? Who possesses sufficient expertise to interpret the significance of possible risks, dangers or hazards? Whose scientific authority will reinforce the impartiality that I am striving to achieve in my news account?
- Wherein lies the news value of a story? Which aspects of it will be of interest to my editor, fellow journalists and audience? Does the story have a clear narrative structure, a straightforward beginning, middle and end, which will allow facts to be expressed quickly and easily?
- What is potentially controversial about this story? If a possible risk is at issue, is it safe? In the event scientific sources refuse to offer a clear-cut 'yes' or 'no' response to my question, how best to communicate possible implications in simple – but not simplistic – terms? How do I put a human face on scientific principles, let alone risk calculations?
- What happens when my sources disagree? If balanced reporting, by definition, means that there are always at least two sides to every story, should alternative accounts be given equal weight in the story? How do I determine what counts as sufficient evidence to sustain a claim in the absence of scientific consensus about its significance?
- How best do I differentiate the public interest from what interests the public?

Looking at this process of mediation from the opposite point of view, namely that of the prospective source, throws an alternative range of possible questions into sharp relief. Scientists, for example, may ask themselves questions such as:

- Why does this journalist want to talk to me? Where will I be made to fit within the larger structure of the news story? That is to say, how will my views be aligned in relation to alternative views, how will the facts of the matter be framed or contextualised?
- Can journalists be trusted to explain complex findings to the public in a manner that will be accurate? Will co-operating with the journalist enhance my profile in the field or will it diminish my credibility, upset my colleagues, antagonise my rivals?
- Will there be benefits for my research in having it turned into a news story? Is this an opportunity to be exploited, especially where attracting potential funding is concerned, or is it best to play safe and pass it up? Do I, as a scientist, have a particular obligation to explain my work to members of the public?
- When discussing controversial science, which may entail possible risks or hazards, how do I acknowledge that it is impossible to eliminate problems entirely without, at the same time, calling into question my authority? In the absence of guarantees, how should I explain which risks should be avoided, which ones may be worth taking, and which ones are insignificant?
- Is scientific uncertainty an inevitable price to pay for progress? Who is entitled to criticise science, and who is not? Wherein lies responsibility for how science is used in society? Is it possible to separate out science from the moral or ethical implications of its application?

Needless to say, the symbiotic nature of the scientist–journalist relationship can quickly become hotly contested under certain circumstances, especially when trust breaks down. It is certainly the case that answers to the questions above will demand attention, as well as careful negotiation, during each and every encounter.

Science on the Web

In rounding out this chapter's discussion, it is worth pausing to consider how science news is changing in the brave new world of the Internet. Across the range of perspectives on offer, it is possible to discern a growing recognition of its importance for redefining what counts as science reporting and – of equal importance – how it will be consumed. Regarding the latter, a recent US study conducted by the Pew Internet Project and the Exploratorium (2006) found that the Internet was a primary source for science news and information (second only to television) for 40 million American adults, and that use of online science resources was 'linked to better attitudes about the role science plays in society and higher assessments of how well they understand science'. More specifically, the November 2006 report's findings, based on a survey of 2000 people in January that year, suggested that:

- Nearly 9 in 10 (87%) online users have used the internet to look up the meaning of a scientific concept, answer a specific science question, learn more about a scientific breakthrough, help complete a school assignment, check the accuracy of a scientific fact, downloaded scientific data, or compare different or opposing scientific theories.
- Most Americans say they would turn to the internet if they needed more information on specific scientific topics. Two-thirds of respondents asked about stem cell research said they would first turn to the internet and 59% asked about climate change said they would first go to the internet. Most of those searches would begin with search engines.
- Nearly three quarters (71%) of internet users say they turn to the internet for science news and information because it is convenient.
- Two-thirds (65%) say they have encountered news and information about science when they have gone online for a different reason in mind.

(Pew/Exploratorium, 2006)

Even bearing in mind the usual sorts of qualifications where opinion surveys are concerned (margins of sampling error – in this study said to be plus or minus three percentage points – as well as interpretations of question wording, practical difficulties and so forth), these results would appear to indicate that the Internet is creating important, and increasingly globalised, spaces for science news and information to circulate. 'People's use of the internet for science information has a lot to do with the internet's convenience as a research tool, but it also connected to people's growing dependence on the internet for information of all types', stated John

B. Horrigan, the study's principal author. 'Many think of the internet as a gigantic encyclopedia on all subjects and this certainly applies to scientific information' (cited in Pew/Exploratorium, 2006; press release 20 November 2006). For science journalists, the report similarly made for interesting reading, not least with regard to its finding that it is a widespread practice amongst users to go online in order to double-check the reliability of science news reports they encounter in more traditional media.

Academic studies into how science news is being interpreted by Internet users are at a relatively early stage of development. Important here is a context-sensitive approach to any bold claim – whether positive or negative – about how people are being influenced (or not) as a result. Using a variety of methods, researchers have sought to discern the subtle, frequently contradictory ways in which public perceptions of science are negotiated as part of everyday life. Rather than invoking a language of causative 'effects' or 'impacts', they strive to understand how people draw upon science news on the Internet to help them make sense of different scientific controversies.[1] Consequently, this type of research invites a reconsideration of certain long-standing assumptions about science news in the Internet age, as well as a heightened degree of self-reflexivity amongst science journalists themselves. Many of the latter, not surprisingly, continue to be sceptical. The Internet – especially in conjunction with the embargo system discussed above – makes it 'easy to churn out story after story, usually without leaving your desk', states Whitehouse (2007), a former BBC science correspondent. 'The result of this', he continues, 'is that science coverage can be indistinguishable across outlets. The quick communication and comparison made possible by the internet has resulted in a uniform blandness of science reporting.'[2]

Speaking in more conceptual terms, Hermida (2007) strikes a more upbeat note, expressing his conviction that the Internet 'offers us new ways of rethinking how science is reported and explained.' As he explains,

> In our traditional print model, we expect audiences to come to us – to pick up the *Vancouver Sun* in the morning and read about the research at UBC [University of British Columbia] into leatherbacks [turtles]. But in an online world, you cannot simply expect people to come to you out of habit. The emphasis is on reaching out

to audiences – to create more opportunities for people to stumble across your content. [...] Increasingly more and more news organizations are taking this approach – providing their audiences with ways of taking their content [to] share it with friends via [the social networking sites] Facebook, YouTube or their own blog.

(Hermida, 2007)

Hermida believes that it is vitally important that science journalists adopt a 'digital mindset', one that retains the values of accuracy and fairness while seeking to capitalise on the non-linear, interactive and participatory nature of social networks in online environments. Chalmers (2007) offers an assessment of the ways in which the Internet is transforming how scientists – in this case physicists – report their findings and communicate with one another. The Internet, he contends, is 'tailor-made' for the 'democratization of science, and is quickly becoming a much more social medium than the one-way "click and download model" of the Web as it was originally conceived'. Blogs such as *Cosmic Variance*, for example, are proving remarkably popular amongst physicists – '3000 readers per day and [it] hosts discussion "threads" ranging from the latest preprints in theoretical particle physics to how to mix the perfect cocktail' – as well as with a wider public of non-scientists (Chalmers, 2007, 20).

Blogs focused on nanotechnology, as one might expect, encompass an extraordinarily diverse range of issues, themes and perspectives. To give a flavour of what is currently available, 'Robots Rule!' (http://www.robotsrule.com/) usefully identifies several of the leading nanotechnology-related blogs as follows:

Nanodot.org
From the Foresight Institute, this blog covers industry news (what company is doing what, sales information, etc.), recent and predicted innovations, legal & legislative related developments and more.

Howard Lovy's NanoBot
Hosted by Blogger Howard Lovy is a true evangelist for nanotechnology. Respected by the Foresight Institute as an expert in his field, his blog covers a variety of nanotechnology related topics including; promoting nanotech education in the schools, job

growth, industry and state economic issues, recent innovations and even the occasional Dilbert comic.

News.NanoApex

More like a nanotechnology oriented news clipping service than a blog, they are a great way to stay on top of all the latest articles and newsfeeds that cover nanotechnology. They also have nanotech discussion forums, a section devoted to micro-electromechanical systems (MEMS), a categorized book list and a small but growing image library.

Roland Piquepaille's Technology Trends

Although technically not a dedicated nanotechnology blog, Roland's blog is often cited by well-respected sites like Slash-Dot.org, a huge number of other tech blogs, and one of my favorite robotics sites Robots.net. Covering the latest advancements in high technology, he is usually one of the first to break a new and usually fascinating story involving the latest in scientific breakthroughs.

Responsible Nanotechnology

A relatively new blog that covers the important topic of the possible positive and negative impacts nanotechnology could have on our planet. The author also covers some of the International political and scientific developments surrounding nanotechnology.

Molecular Torch

This blog is dedicated to following the state of the art in the rapidly emerging field of nanocrystals, nanoparticles, or quantum dots.

(http://www.robotsrule.com/html/nanotechnology-blogs.php)

Taken together, these blogs offer a general indication of some of the broad features of 'nano-blogging' at a time when its conventions are gradually becoming consolidated. Simply typing 'nanotechnology blogs' into Google.com will generate a dazzling number of sites. Many of them, upon close inspection, prove to be useful resources for journalists and their audiences alike, while others strive to advance particular stakeholder interests or viewpoints under the guise of impartial news. Each and every blog, needless to say, should be read with a sceptical eye.

Precisely how, and to what extent, the Internet is changing the characteristics of science news is fast becoming a research topic in its own right, especially in relation to emergent sciences such as nanotechnology. Bowman and Hodge (2007) contend, for example, that there is a serious 'knowledge gap' where this 'scientifically complex and often overhyped' technology is perceived. Hence their fear that 'a failure to engage citizens in proactive dialogue activities may create the perception that nanotechnology products have little legitimacy in our society and economy, and may therefore result in a consumer backlash' (2007, 128). Bearing in mind related examples of emergent technologies – the introduction of genetically modified organisms (GMOs) in crops and food products being an obvious example – where public dialogue problems became critical, it is vital to learn from past mistakes. To the extent that science news can draw upon the interactive resources available via the Internet, then, it may help to enrich and deepen the conditions for deliberation and debate. 'With sound technical data about nano-materials' health and environmental impacts and a commitment to open dialogue about potential social and ethical implications with all stakeholders', Kulinowski (2004) believes, 'nanotechnology could avoid traveling along the wow-to-yuck trajectory' (2004, 19). In any case, even those who are dismissive of the more optimistic claims being made about the web's potential for creating new spaces for 'public engagement with science' need to recognise that it promises to dramatically recast the news reporting of nanotechnology in the years to come.

To close, this chapter has sought to provide an overview of a range of issues central to current debates about the reporting of science. Singled out for scrutiny have been certain key issues concerning the intense pressures constraining the reporting process, journalistic perceptions of the relative authority of informational sources and the framing of scientific concepts and principles, amongst others. The recurrently routine, everyday way in which science is rendered newsworthy (or not) has been shown to entail a complex process of negotiation, one performed by journalists in relation to an array of factors. Given the extent to which decisions about these factors are typically regarded by journalists to be a matter of 'common sense', special attention has been devoted to unravelling the largely unspoken, taken-for-granted conventions that underpin these seemingly *ad hoc* judgements. In this regard, the dynamics of journalist–source relationships have been shown to be of particular importance (a point explored more substantively with regard to nanotechnology in the chapters to follow). Precisely how these relationships continue to evolve – and, equally important, how they may be improved in time – raise challenging questions, especially where the Internet is concerned. There is little doubt that what will count as science reporting in the years to come will be profoundly influenced by the web, where new forms of journalism such as blogs are recasting familiar norms and inviting new types of interactivity with audiences. The implications for the news reporting of nanotechnology, we have argued, warrant very close attention indeed.

Beginning in the next chapter, our discussion turns to the pressing need to better understand how the media represent risk, in general, and the possible dangers, threats and hazards engendered by nanotechnology, in particular. Drawing upon a number of theoretical models, including Beck's (1992a) conception of the 'risk society', we seek to discern the ways in which the news media help to shape the 'relations of definition' – to use Beck's phrase – framing public controversies about nanotechnology and its applications.

Notes

1. Relevant examples of research undertaken to date include Eveland and Dunwoody's (1998) examination of Internet users' engagement with 'The Why Files' website, which aims to provide 'the science behind the news' in story narratives. Richardson (2001) critiques the dynamics of news exchange via Internet newsgroups discussing possible health risks of BSE, or 'Mad Cow Disease'. Eisend (2002) investigates the effects of Internet use on traditional scientific communication media among German social scientists. Treise et al. (2003) offer insights into the factors appearing to influence the perceived credibility of a science website in the US. Koolstra et al. (2006) compare Dutch television with the Internet in terms of public perceptions of their relative reliability as sources of information for science communication. Weigold and Treise's (2004) study examines how teenagers use the Web to find science information, suggesting that participants 'regularly link from news stories to science sites while reading interesting science-related news stories' (2004, 237).

2. Pertinent here is Trumbo et al.'s (2001) investigation of email and Web use by science journalists in the US, suggesting that the rapid diffusion of these technologies has been due, in the main, to a positive orientation towards the quality of Web information, trust in its sources and advantages

gained through connectivity. Massoli (2007) examines the 'journalistic information approach' adopted on the websites of various European public research institutions. Pinholster and C. O'Malley's (2006) study focuses on how EurekAlert!, the science news Web service of the American Association for the Advancement of Science (AAAS), helps reporters to cover the fast-breaking science beat around the globe.

3
Risk Society and the Media

Science is increasingly exposed to contestation and criticism within contemporary society as new forms of risk are seen to emerge and are amplified by the media. Familiar questions about how the news media frame the significance of emerging technologies are being re-cast by nanotechnologies. Precisely how the parameters of debate are defined will be crucial for future policy developments and very likely for how publics react. Concerns about the risks associated with nanoparticles, for example, could play a part in stigmatising nanotechnology – a condition that is very hard to dislodge once it takes root. Where there is considerable indeterminacy and ambiguity in a scientific field, the media have the potential to play a gatekeeping role through deciding who becomes a primary definer of the issues and whose views are presented as having the greatest legitimacy.

This chapter situates the media politics of nanotechnology within broader debates concerning the role of the news media in communicating risk. Given the extensive literature on media, risk and science, we ask what novel issues does nanotechnology raise about the nature of risk reporting? We consider what constitutes 'newsworthiness' and credible news sources as well as the processes through which particular themes and voices become dominant or marginalised. In particular, we examine the question of why there has been an absence of reference to particular actors and issues, such as economic, ethical, theological, and legal issues, and whether a particular framing is contingent on certain events or periods in the policy-making cycle. In the discussion that follows we argue that in the case of nanotechnologies the news media are especially likely to be a hot house of definitional debates. At this early stage in the issue attention cycle we suggest that 'risk' is more open to contention since currently there is little public awareness of the issues, and the language and metaphors have yet to be settled. We conclude by considering how we can move beyond earlier approaches to provide a more sophisticated account of the shifting role of the news media in communicating risks within contemporary society.

Risk, uncertainty and social values

Problematising risk

The media have a crucial responsibility for communicating information to publics about scientific issues within contemporary society (Anderson, 2006; Hargreaves and Ferguson, 2000; Kitzinger et al., 2003). As we saw in Chapter 1, complex scientific debates over issues such as GM food and crops, stem cells, climate change and nanotechnologies present us with new challenges. Over recent decades there has been heightened interest in media and risk, both within the news media and in scholarly publications (Anderson, 2006; Gurabardhi et al., 2005; Kitzinger, 1999; Mythen, 2004). Researchers examining the relationship between risk and the media have to deal with the thorny issue of how to define 'risk' (Allan et al., 2005; Kitzinger, 1999). The category of 'risk' is extremely wide and embraces a diverse range of issues. Moreover, risk reporting straddles many different fields (environment, science and technology, war reporting, consumer issues and health coverage, to name but a few) and does not map neatly onto established journalistic beats.

The traditional approach within the risk communication literature judged risk reporting in terms of the degree to which it could be seen as an accurate reflection of expert opinion. Also there was a widely held assumption that science needs to be 'popularised' because of a lack of public 'understanding' or knowledge. However, this 'information deficit' model has come under increasing attack. The problem was seen to simply stem from an 'information vacuum' between scientific experts and lay publics. As Gregory and Miller observe, 'Within this model . . . scientists are the providers of all knowledge, and the arbiters of just what should be provided, to an empty-headed public' (2001, 61). Being aware that they are never completely in control of how their views are framed, scientists themselves often express concerns that the media 'hype', 'distort' and 'misrepresent' their research, thereby reinforcing public mistrust in science and creating a backlash. As we shall see in Chapter 4, the scientists interviewed in our ESRC-funded study were sharply critical of the sensationalist nature of the coverage that nanotechnologies were attracting. Interestingly, these views are broadly typical of attitudes towards science coverage in general (Allan, 2002; Miller, 2001; Nelkin, 1995).

What makes nanotechnology unique?

Given the burgeoning literature on science, risk and the media (e.g. Allan, 2002; Bauer and Bucchi, 2007; Bauer and Gaskell, 2002; Brossard et al., 2007; Nelkin, 1995) what new questions, if any, does nanotechnology raise about the nature of risk reporting? There are a number of parallels that can be drawn between nanotech and past controversies over other 'emerging technologies' such as biotechnology and stem cell research. The hype itself is nothing new; the heightened expectations surrounding nanotechnologies are similar to those that accompanied the emergence of biotechnology (Berube, 2006; Brown, 2003; Bubela and Caulfield, 2004; Jensen, 2008; Jones, 2008; Marks et al., 2007). However, from the start a concern with societal implications has been at the forefront of nanotechnology development, given the fears that a GM-style public backlash may occur. Indeed, evidence suggests that negative frames have featured more heavily in early press coverage of nanotechnologies, especially in the UK, compared with the first two decades of biotechnology coverage (Anderson et al., 2005; Gorss and Lewenstein, 2005). Despite this, in overall terms the degree of importance placed on economic progress has been particularly prominent especially in the US. The emphasis from the beginning upon societal benefit and consumer applications, itself part of a larger movement towards encouraging scientists to produce more commercially exploitable products, can be seen, to some extent, to distinguish it from other scientific developments (Gorss and Lewenstein, 2005).

Increasing public relations (PR) pressure often leads to hope turning into hype and simply blaming the news media for sensationalism ignores the bigger picture (Kitzinger, 2006). The hype surrounding nanotechnologies reflects, to a degree, increasing pressure from funding agencies to promise research that has clear economic impacts (Jones, 2008). Expectations play a key part in establishing legitimacies, defining roles and mobilising interest and resources, especially during the early phases of technological developments (see Anderson, 2007; Brown, 2006; van Merkerk and Robinson, 2006). As Selin observes,

> expectations serve a very real, very palatable role in the development of nanotechnology... That is, the future is a rhetorical and symbolic space to work out "what is nanotechnology?", but also serves a productive role that underlies everyday decision making, alliance building, and resource allocation. Potentiality is shifted and revised based on the agendas, interests and needs of those engaged in the space. Both as a rhetorical and blatant theoretical chartering of nanotechnology, potentiality is used and manipulated by various actors to gain and lose allies, muster authority, and legitimate projects.
>
> (2007, 214–5)

An abundance of hope has come to be associated with nanotechnology in a whole range of different spheres of life. Biotechnologies have provoked controversy because they involve control over life itself. Cloning, pre-implantation genetic diagnosis and embryonic stem cell research are all seen as challenging ideas of the natural, when life begins and ends, and concepts of identity. However, nanotechnologies *also* include applications involving control over life (nano-medicine) but in addition find applications in consumer products [e.g. sunscreens, face creams, clothing, sports items (tennis balls)] as well as electronics and engineering, environmental sustainability, energy and the military. In the US, a significant amount of federal funding is allocated to military investment in nanotechnology R&D, to advance offensive and defensive military objectives (e.g. creating battle-suits that give soldiers better protection in combat, nanosensors and nanoelectronics) (Kulinowski, 2006, 18). This raises the possibility that a controversy in one field of application may 'spill over' into other fields of application, heightening anxiety over potential risks and, in the process, weakening trust in institutions and regulatory processes governing that field. Constructions of risk cannot be easily contained within particular technological fields and, for risk managers, there is the continual danger that perceptions of one technological development could be 'contaminated' by perceptions of another. The predicted future convergence of technologies, such as nanotechnologies, biotechnologies and information and communication technologies, is likely to lead to new constructions and perceptions of risk (see Petersen et al., 2008).

Clearly the scale of nanotechnology means the risk is 'invisible' with any contamination indeterminable to our sensory capacities (Kasperson et al., 2001, 16). Thus far there has been relatively little research into health and safety aspects of nanotechnologies, and even less work has been conducted into environmental risks. This, together with the technologies' unbounded character, increases the potential for alarm. Some evidence suggests, however, that the risk perception dynamics are more complex with nanotechnology compared to previous controversies concerning emerging technologies. A recent study by Scheufele et al. (2007) found that nano-scientists were generally more optimistic than the public about predicted benefits yet for certain issues, concerning potential pollution and longer range health risks, they were considerably more concerned. Given the level of indeterminacy about its 'real' or 'imagined' effects, truth claims surrounding nanotechnology have become highly contested within the administrative policy arena. Furthermore the relationship between science and society is changing, increasingly dominated by the management of trust and the politics of spin, despite the new emphasis upon 'upstream' public engagement (see Chapter 5). 'Engagement' itself can function as a type of risk management; a means of engineering consent and legitimacy for projects and technologies for which there is considerable scientific and government backing, and which are seen as potentially placed in jeopardy by an untrusting and resistive 'public'. In this discourse 'the public' tends to be implicitly perceived as unitary, rather than made up of heterogeneous constituencies with varied views on nanotechnologies derived from a range of sources and with differing interests and degrees of ability and propensity to 'engage' with the issues (see Petersen et al., 2008). Accordingly, new questions arise about the role of the news media in reporting such 'invisible' risks within a culture where science journalism tends to be source-driven and source-framed.

Social factors in the production of news

Only a limited number of studies have focused upon news media coverage of nanotechnologies and their findings are largely tentative, given that we are at an early stage of issue development (see Anderson et al., 2005; Faber, 2006; Friedman and Egolf, 2005; Gaskell et al., 2005; Gorss and Lewenstein, 2005; Lewenstein et al., 2004; Stephens, 2004, 2005; Te Kulve, 2006). As we saw in Chapter 2, the concept of framing refers to principles of selection based on core values and assumptions that inform a journalist's interpretation of a news event. In the case of nanotechnology, the dominant framing has yet to become established. Amongst those studies where framing issues are analysed, it is found that news coverage – both in the UK (see Anderson et al., 2005) and the US (see Gaskell et al., 2005) – tends to be concentrated within a relatively small number of elite newspapers. As we shall see in Chapters 5 and 6, it is clear that scientists' perceptions of the role of the newspaper press may vary considerably, reflecting differing assumptions about the nature of science and the role of the scientist, about 'the public' and 'its' knowledge and motivators to action, and concerning the relative significance of other media and non-media influences (e.g. personal backgrounds and experiences) on the formation of public opinion (Hansen, 1994; Gunter et al., 1999). It can be anticipated that scientists' familiarity with previous controversies, such as those surrounding BSE, GM crops, cloning and stem cell research, which have attracted considerable news media coverage, will necessarily influence their expectations about how future developments will develop – for better or otherwise (Allan, 2002; Gibson, 2003). Similarly, we argue that journalists' previous experiences of covering such science controversies are likely to shape their treatment of nanotechnology. Early UK newspaper coverage of GM food and crops, for example, was found to parallel earlier reporting of BSE (Murdock, 2004). As we shall see in Chapter 4, the journalists interviewed in our ESRC study tended to refer to these previous, high-profile cases.

News media coverage of risk does not simply mirror reality; it is highly selective and subject to a range of social and cultural influences. Weighing up the degree to which media representation of risks is 'accurate' and 'balanced' is extremely complicated and far from straight forward. However, those who accuse the media of routine sensationalism tend to assume that official sources operate purely on a scientific basis. As Kitzinger (1999, 62) observes.'... the important questions are not to do with whether the media "play up" or "play down" risk but *which* risks attract attention, *how, when, why* and *under what conditions?* (original emphasis)'. Some risk practitioners tend to naïvely assume that the media will automatically cover risk issues but, as Eldridge and Reilly (2003, 139) point out, the news media are poorly adapted for sustaining high-profile ongoing reporting of long-term risks, especially when there is great scientific uncertainty and official sources remain largely silent. Journalists do not generally like dealing with scientific uncertainty (see Anderson, 1997; Friedman et al., 1999). Indeed, there is a fundamental conflict between the journalistic attraction to conclusive or controversial new findings and the tendency for scientists to qualify statements. Moreover, the concept of 'risk' deals with speculation about what may happen a long way into the future and is in conflict with news schedules that emphasise current events. Such potential risks are likely to be simply ignored unless they are manifesting themselves in some concrete way. The 'invisible' size of nanoparticles means that the topic often has little journalistic appeal unless it can be associated with aspects of popular culture that are seen to have strong resonance; for example, Michael Crichton's novel *Prey*, or the comments of a key public figure such as Prince Charles. In other instances, rumours proliferate in the news media about supposed risks linked with new technologies; a case in point being periodic panics over the potential health effects of mobile phones (Burgess, 2004).

It is not necessarily the most serious risks that command the most attention. As Kasperson et al. observe, 'The mass media cover risks selectively, according those that are rare or dramatic – that is, that have "story value" – disproportionate coverage while downplaying, or attenuating, more commonplace but often more serious risks' (2001, 18). Coverage tends to be event-driven and thrives upon unexpected, dramatic disasters or unusual risks (see Anderson, 1997; Kitzinger, 1999). The 'human interest' factor is often an important ingredient in news stories on risk. Commercial pressures encourage the news media to focus upon issues and themes that are likely to be meaningful and familiar to their audiences (Allan, 2002; Hansen, 1993; Kitzinger and Reilly, 1997; Williams et al., 2003). This is particularly problematic for nanotechnology given that there is currently little public awareness about its applications and potential effects (Bainbridge, 2002; Cobb and Macoubrie, 2004; Gaskell et al., 2005; Renn and Roco, 2006; Waldron et al., 2006). The search for dramatic stories leads media practitioners to focus upon 'breaking news' and discourages covering issues that slowly unfold over time. Risks that lead to the death or serious illness of large numbers of people in a single event – such as nuclear accidents – are more likely to attract coverage than those which have a gradual cumulative impact over a lengthy period – such as those connected with asbestos exposure or smoking (Greenberg et al., 1989; Singer and Endreny, 1987). However, a number of studies have found that in general the tone of risk reporting tends to offer reassurance rather than signal alarm (Petts et al., 2001; Schanne and Meier, 1992). Indeed, the accusation that media coverage is sensationalist itself is likely to stem from selective memory (Kitzinger, 1999).

Risk as a political issue

Where there is considerable scientific uncertainty over risk it is to the news media that 'ordinary' or 'lay' members of the public are likely to turn in order to gain a better grasp of the issues. Media practitioners are expected to render the underlying significance of uncertainties intelligible for their audiences' everyday experiences of contemporary life. News accounts tend to offer the assurance that a potential risk will remain uncertain only until further research and scientific investigation are able to provide the expected certitude and clarity (see also Adam, 2000a). In this context, risk becomes a deeply political issue. As Hornig observes, 'In a post-industrial democracy confronting the social acceptability of the risks of emerging technologies is an everyday form of crisis' (1993, 95). In examining the decisive influence of the media in shaping public debate over risk policy issues, Hornig highlights the ways in which diverse political actors use scientific opinion to advance their interests. She draws attention to two distinctive approaches to the evaluation of risks, referred to as the 'rationalist' and the 'subjectivist' perspectives. These are underpinned by a very different range of assumptions concerning scientific opinion as to how different levels of risk are best determined, the ideal role of the media in representing risk and whether objectivity is viewed as possible or even desirable (Anderson, 1997; Lichtenberg and MacClean, 1991). According to the rationalist approach '...it is theoretically possible, if sufficient data could be collected and various technical problems of analysis solved, to arrive at an absolute measure of the riskiness associated with any technological innovation'

(Hornig, 1993, 96). These data can then be used as a kind of yardstick against which the sway of public opinion and media representations can be calculated. The solution, according to this approach, is to enhance risk-related decision-making by making sure that systematic 'distortions' are recognised and set right through educating publics. From this perspective, '... media accounts of risk are typically judged on how accurately they reflect the scientific point of view and how well they contribute to public education designed to eradicate wrong thinking' (Hornig, 1993, 96).

By sharp contrast the 'subjectivist' position places emphasis upon the socially constructed nature of risk. Various psychometric studies of risk perceptions highlight that '... the evaluation of risk information takes place in a social context and involves value judgements and priorities – that is, that this process is inherently subjective' (Hornig, 1993, 96). This approach, therefore, focuses on the social processes in and through which particular definitions of risk are produced. In particular, it draws attention to the ways in which contending definitions may be mobilised to lend credence to a certain understanding of the risk issue in question. As Hornig observes, the media play a key role in this process as '... they vocalize and therefore legitimize some points of view (often those of established institutional news sources) and ignore others' (1993, 96). This can clearly be seen in the way in which the UK national press framed nanotechnologies during 2003 and 2004 – the formative period in their rising public salience. As discussed further in Chapter 4, the findings of our ESRC-funded study reveal how the views of critics (e.g. Prince Charles, ETC group) and advocates (e.g. Tony Blair, Lord Sainsbury) were positioned in a way that accentuated their differing viewpoints. Prince Charles's public comments on nanotechnology tended to be portrayed as ignorant and scaremongering and firmly placed with the antiscientific views of the 'nano-luddites' (see Anderson et al., 2005). In this way the news media frame certain truth claims as reasonable and credible whilst others are ignored, trivialised or marginalised. Yet both the rationalist and the subjectivist positions view risk reporting as problematic, though from very different vantage points. While rationalists often blame the media for distorting and politicising the 'objective facts', subjectivists frequently attack the media for over-representing the scientific point of view in positive terms.

Having underscored the importance of recognising that interpretations of 'risk' are underpinned by different value judgements and priorities, we now turn to briefly examine risk society and the psychometric paradigm. First we discuss how each can be seen to shed light on the unfolding of the nanotechnology debate and, following this, we critically reflect upon their limitations.

Risk society and mediating the limits of risk

Of particular relevance here is the work of Ulrich Beck and his 'Risk Society' perspective which highlights the uncertainty and unpredictability of contemporary risks. Central to his thesis is the shifting nature of hazards in 'pre-industrial', 'industrial' and 'risk' societies. Beck contends that 'manufactured risks' in contemporary society (such as those associated with nanotechnologies) are more catastrophic than 'natural hazards' in pre-industrial society (such as earthquakes, plagues and seasonal droughts), because they are no longer geographically and temporally contained and they escape established welfare and security systems. The humanly produced 'manufactured risks' have unpredictable effects that radiate far beyond a specific geographical locale, or a particular point in time. For example, the potential military applications of nanotechnologies could, he argues, lead to the creation of 'dirty' bombs; the effects of which would be felt well beyond the immediate location and suffered by generations to come (Beck, 2002, 2006). Thus Beck claims,

> The modern world increases the worlds of difference between the language of calculable risks in which we think and act and the world of non-calculable uncertainty that we create with the same speed of its technological developments. With the past decisions on nuclear energy and our contemporary decisions on the use of genetic technology, human genetics, nanotechnology, computer sciences and so forth, we set off unpredictable, uncontrollable and incommunicable consequences that endanger life on earth.
>
> (2002, 3)

The riskiness of everyday life is becoming increasingly apparent as uncertainties deepen. As we become more reliant on 'experts' to manage and control risks, our trust in these same experts continues to lessen. Paradoxically, the risk society produces and tries to legitimise the very hazards which are beyond the control of its institutions. As Beck observes,

> The entry into risk society occurs at the moment when the hazards which are now decided and consequently produced by society *undermine and/or cancel the established safety systems of the provident state's existing risk calculations*. In contrast to early industrial risks, nuclear, chemical, ecological and genetic engineering risks a) can be limited in terms of neither time nor place, b) are not accountable according to the established rules of causality, blame and liability, and c) cannot be compensated or insured against. Or, to express it by reference to a single example: the injured of Chernobyl are today, years after the catastrophe, not even all *born* yet.
>
> (1996, 31, original emphasis)

Thus the transition to manufactured risks produces a crisis of responsibility. The precautionary principle is often advocated, as is the case with nanotechnologies, in order to get companies to think through the whole product cycle before goods are released onto the market (see Throne-Holst and Stø, 2008). In this context of heightened uncertainty the media play a crucial role in the formation of public understandings of risk. This is especially so with nanotechnology given the degree of speculation that exists over potential effects. 'Risk' itself becomes more open to contention given the low public visibility of the issues and the fact that the language and metaphors through which the science is communicated have yet to become established.

In *Risk Society* – first published in 1986 – and in a series of later publications, Beck argues that in late modernity the media can act as a vehicle through which experts relay institutional information to publics and, paradoxically, as a channel for reflexivity and public critique. Particularly pertinent here is his concept of 'relations of definition' underpinning media discourses which condition what can and should be said about risks, threats and hazards by 'experts' and 'counter-experts', as well as by members of the 'lay public'. The analysis of risk, he claims, needs to account for the media's structurating significance in the formation of public opinions about risk. We are dependent upon the media to comprehend the 'world out there' beyond our immediate experience. Important questions therefore arise concerning who in the media wields this spotlight, under what circumstances and, moreover, where it is (and is not) directed and why. The role of the media in representing issues of proof, account-ability and compensation becomes of crucial significance. Risks, as Beck (1992a, 23) suggests, can 'be changed, magnified, dramatized or minimized within knowledge, and to that extent they are particularly *open to social definition and construction*' (original emphasis). The ways in which journalists help to mediate the limits of risk, both in con-junction with and opposition to other social institutions, therefore need to be carefully examined.

News media coverage of nanotechnologies is inevitably selective (as discussed in Chapter 2), reflecting numerous factors such as news values; time and space constraints; the nature and type of social relations that journalists develop with their sources; advertis-ing pressure; ownership and ideological cultures; editorial policies; a journalist's own personal knowledge of the area and perspective on it; and the extent of policy activity in the field (Allan, 2002; Anderson, 1997). A range of studies focusing on previous biotechnol-ogy controversies underscore that particular framings are contingent upon certain events and periods in the policy-making cycle (Gorss and Lewenstein, 2005; Nisbet and Lewenstein, 2002; Nisbet et al., 2003). Journalists tend to be guided by the ideal of objectivity, along with professional commitments to balance and fairness, lead-ing to an emphasis on verifying the 'facts' often through approaching authoritative sources before looking for conflicting views. Part of the professional orientation of journalists is to cultivate 'trustwor-thy', 'credible' and 'legitimate' sources, mindful of the pressures to safeguard their own integrity as reporters. This 'strategic ritual of objectivity' (Tuchman, 1978) tends to advantage powerful institu-tional sources (Stocking, 1999). Media practitioners routinely gain material from a range of sources including press releases, press con-ferences, PR officers, professional society meetings, scientific journals and interviews (Nelkin, 1995, 105). However, they frequently lack the time, means or expertise to search out independent verification of facts (Dunwoody, 1999; Miller and Riechert, 2000), sometimes being over dependent on pre-packaged information (Goodell, 1986; Logan, 1991; Manning, 2001; Petersen, 2001). The sheer complexity of the nanotechnologies field, which cuts across a wide variety of scientific disciplines, poses particularly acute problems for reporters. As we explore further in Chapter 4, the science correspondents interviewed in our study expressed concern over the validity of the term 'nanotechnology' and the extent to which it constitutes an identifiable field.

As sources, scientists potentially wield a considerable degree of control over the news production process. Indeed, a significant amount of material is source generated – some estimate it may account for as many as half or more newspaper stories, so scientists are able to strategically package their material for news media (Nisbet and Lewenstein, 2002, 362). A tendency to accept press releases on their own terms can often lead general reporters, in particular, to produce exaggerated claims – a problem acknowledged in the development of nanotechnology communication guidelines for reporters in the US (Bell, 2006). Although journalists may choose the topic for a news item, scientists often aid in defining the boundaries from which story choices are made (Dunwoody, 1999, 63). Industry also exerts a significant influence on the news reporting of nanotechnology and lessons from previous controversies over emergent technologies highlight the importance of attending to public perceptions (Meili, 2005). In the nanotechnologies field, industry sources have gained considerable media coverage, especially in the US, and PR is increasingly prominent (Cooper and Ebeling, 2007; Stephens, 2005).

Our analysis suggests it is important to identify why there has been an absence of reference to particular actors and issues, such as economic, ethical, theological and legal issues, and whether a particular framing is contingent on certain events or periods in the policy-making cycle. Research suggests that elite US national newspapers have tended to treat nanotechnology principally as a business story emphasising scientific discovery and progress, whilst UK national newspapers have given comparably greater attention to social, legal and ethical issues (Gorss and Lewenstein, 2005; Stephens, 2005). As we explore further in Chapter 4, NGOs did not feature prominently as quoted or cited sources in our study. The analysis of previous science controversies (such as stem cell and climate change research) demonstrates how scientists tend to be the principal news sources during the early stages when an issue is located within an administrative policy arena (Carvalho and Burgess, 2005; Nisbet et al., 2003). Less-advantaged groups may seek to shift the framing of the debate away from narrow technical terms by creating 'pseudo events' that dramatise the issues and associate them with emotionally laden symbols (Anderson, 1997; Gorss and Lewenstein, 2005). Media attention does not typically peak until competing interests force the expansion of the issue and it moves to more overtly political contexts (Carvalho and Burgess, 2005; Nisbet et al., 2003).

It is important, therefore, to avoid overlooking the ways in which scientists, and other stakeholders, may attempt to control news at various stages of its production. This may occur not only through the regulation of the flow of information (e.g. news conferences, press releases, 'leaks'), but also through the promotion of preferred imagery, language and rhetorical devices. Also, assumptions concerning the separation between 'science' and 'popularization' need to be challenged, especially when it denies the input of popular views into the research process and the simplification that is an intrinsic aspect of scientific communication (Hilgartner, 1990, 523–4; Lewenstein, 1995). Journalists may often be specialists in their field, some of whom have a scientific training.

The identification of the fissures, slips, silences and gaps in media reporting needs to be accompanied at the same time by a search for alternatives. New ways need to be discovered to enhance science journalism and equip it to deal with moral and ethical responsibilities that respond appropriately to emergent technologies such as nanotechnology. Beck's work draws attention to the key role played by the media in articulating competing rationality claims. In seeking to de-naturalise the ways in which the media present certain actors as being self-evidently 'expert' and 'authoritative' while simultaneously framing others as lacking 'credibility' and 'authority', it is the very self-evident nature of this which needs to be questioned. As we shall see in Chapter 4, media representations are far from an 'objective' account of reality; they represent a contested discursive terrain. Irrespective of what some journalists might claim, facts do not simply 'speak for themselves', and actors calling into question scientific rationalities are not, by definition, 'misinformed', 'irrational' or 'anti-science' (see also Adam et al., 2000; Cottle, 1998; Macnaghten and Urry, 1998).

The risk society thesis is a universal, macro-level theory that paints a broad-brush picture of social change which is lacking in empirical detail and sweeps over the complexities of modern life. However, in order to assess its contribution we must consider its purpose. As Mythen contends, 'The risk society thesis is assembled in the spirit of exploration and adventure: it is not driven by empirical validity, but by invigorating sociology and providing thought-provoking reflections on the modern condition' (2007, 803). This needs to be borne in mind when considering some of its deficiencies. While Beck's work has undoubtedly made an enormous contribution towards understanding the nature of contemporary society, it has been critiqued as over simplistic, unevenly developed and in places contradictory (Cottle, 1998, 25; Mythen, 2004, 76). Before turning to reflect more broadly on the usefulness of the risk society theory for an understanding of the role of the news media in representing nanotechnology, we briefly consider the contribution of psychometric approaches.

Social amplification

The Social Amplification of Risk Framework (SARF) was developed by Kasperson and colleagues in 1988 with the aim of producing an integrative conceptual framework for a fragmented range of risk perspectives in the expanding field of risk communication and perception (see Kasperson et al., 1988; Pidgeon et al., 2003; Renn et al., 1992). This new framework sought to bring together insights from a diverse array of approaches in studies focusing on psychological, media, cultural and organisational aspects of risk. It set out to explain why particular risk events, defined by experts as non-threatening, attract significant socio-political attention (amplification), while other risk events, judged to pose a larger objective threat, gain little prominence (attenuation). According to Pidgeon et al. (2003, 2) the SARF

> aims to examine broadly, and in social and historical context, how risk and risk events interact with psychological, social, institutional and cultural processes in ways that amplify or attenuate risk perceptions and concerns, and thereby shape risk behaviour, influence institutional processes, and effect risk consequences.

The original model identified four principal pathways or mechanisms through which risk amplification takes place: signal value, heuristics, social group relationships and stigmatisation (Kasperson et al., 1988). Signal value refers to what a risk event signals or warns. Heuristics are short-hand mechanisms that people use everyday in order to evaluate complex risk information. The character of the social relationships within social and political groups (e.g. advocacy groups, industry, government agencies) affect how risk 'mutates' and publics respond. Finally, stigmatisation, a concept originally applied to marginalised individuals and groups, designates a situation where a technology becomes 'tainted' or 'blemished' by discourses of risk, and is extremely hard to overcome once it takes root. As Gregory et al. observe,

> Stigma goes beyond conceptions of hazard. It refers to something that is to be shunned or avoided not just because it is dangerous but because it overturns or destroys a positive condition; what was or should be something good is now marked as blemished or tainted... Technological stigmatization is a powerful component of public opposition to many proposed new technologies, products and facilities.
>
> (2001, 3)

Indeed, recent work has highlighted the importance of stigmatisation as a key mechanism through which the amplification of risks can produce ripple effects (such as regulatory demands) well beyond the initial impact (Flynn et al., 2001). In the case of nanoparticles the possible impetus for stigmatisation is clear, particularly given that the science involves many of the characteristics which are frequently found in stigmatised conditions (see Wilkinson et al., 2007). Humans are already exposed to natural sources of nanoparticles through, for example, the natural geological process of weathering. However, they are also likely to be exposed to engineered nanoparticles through the air, food or water supply, or through cosmetics or medical products (Handy and Shaw, 2007). The environmental concentrations and toxic effects via each human exposure route are very hard to measure and are currently unknown. However, preliminary findings from toxicology studies suggest the possibility of damage to immune systems, respiratory health and the potential for carcinogenic effects (Handy and Shaw, 2007; Hannah and Thompson, 2008). Moreover, the scale of the technology means the risks are 'invisible' with any contamination indeterminable to sensory capacities (Kasperson et al., 2001,16). Nanomaterials can enter through the intestinal tract, the lungs and the skin, although skin presents a more complex barrier for nanoparticles to penetrate (Hoet et al., 2004). Despite limited exposure to nanoparticles at present, and the toxicology of nanoparticles being informed by lengthy experience dealing with particle safety, they are 'superficially familiar' risks since engineered nanomaterials differ significantly from the particulates traditionally examined (Colvin, 2003, 1167).

While stigma is a useful concept, one of the problems with SARF is that it tends to assume that the process begins with an actual, physical 'risk' that exists before social institutions react, which increases or decreases depending upon the activity of these institutions (see Hornig Priest, 2005). In the next section, therefore, we begin to tease out some of the complexities of the process whereby issues become defined as 'risks'.

Fields of cultural production

Like Beck, SARF theorists have tended to produce a media-centric reading of news-media framings. This tends to be underpinned by pluralist assumptions that news sources are competing on the same terms rather than possessing differing degrees of cultural capital. Social institutions differ in their ability to influence news media agendas and the wider policy-making arena. As Murdock et al. (2003, 162) maintain, '...work conducted within the SARF framework tends to concentrate on the amounts of coverage given to different issues and actors rather than the terms on which this publicity is secured'. They provide a more sophisticated account of source–media interaction, highlighting that control over the media is as much about the power to suppress or silence issues as it is to publicise them. Murdock et al.'s (2003) alternative model adapts Bourdieu's (1998) concept of 'fields of action'. Here risk is usefully viewed as a field of contest where news sources are involved in constant definitional struggles. Accordingly, this requires a multi-faceted analysis of the factors affecting the effectiveness of media strategies over time.

As Cottle (1998) notes, the media-centric focus of Beck's analysis also directs attention away from analysing processes of news production and is 'conspicuously silent... on the institutional field in which "relations of definitions" compete for public recognition and legitimation' (Cottle, 1998, 18). This gap needs to be addressed though systematic empirical analysis of the relationship between news sources and the media, and theorisation that is able to effectively explain the complexity of media involvement in framing risk (Anderson, 1997; Cottle, 1998). Previous research demonstrates a more complex understanding of source–media relations than is often portrayed in the literature (see Anderson, 1997, 2003; Kitzinger and Reilly, 1997; Kitzinger et al., 2003; Manning, 2001; Miller and Riechert, 2000). Although countless studies have highlighted that official sources tend to dominate risk coverage, especially where a major crisis is concerned (see Anderson, 1997; Kitzinger, 1999; Schanne and Meier, 1992; Stallings, 1990; Williams et al., 2003), elite access is not automatically guaranteed. News entry is dependent on numerous factors (internal and external to the media) and research demonstrates that under certain circumstances marginal groups can gain extensive news access (e.g. Anderson, 1997; Hansen, 2000; Manning, 2001). However, achieving access is only half the battle; how news sources' claims are framed, and whether they are treated as credible and legitimate, is of critical importance. As Ryan (1991, 53) observes, 'the real battle is over whose interpretation, whose framing of reality, gets the floor'. In this regard the rise of the PR industry is especially significant, as news sources make increasing use of PR practitioners to package their claims (see Cottle, 2003; Davis, 2000; Miller and Dinan, 2000). However, the competition to gain favourable coverage does not commence from a level playing field, since official sources tend to have greater financial leverage and abundant stocks of cultural capital at their disposal (Anderson, 1997). While non-dominant sources may lack the finance, status and PR personnel advantages enjoyed by official sources, they are sometimes able to react more rapidly to media demands because they are not so constrained by burdensome bureaucratic processes and/or political restrictions (Anderson, 1997; Kitzinger, 1999; Kitzinger and Reilly, 1997).

   A further problem with SARF, as with the risk society thesis, is the tendency to display a simplistic grasp of the influence of media upon policymakers and the public. SARF assumes a linear flow of messages, based on an outmoded transmission model of media effects, founded on the notion that the problems stem from the distortion of expert knowledge whilst being transmitted to publics which leads to exaggerated or false perceptions (Hornig Priest, 2005). Instead Murdock et al. (2003, 161) see risk as a field of contest where claims makers compete with one another to frame issues and seek to influence public opinion. As Petts et al. (2001, x) argue,

> SARF at best provides a highly simplistic understanding of the role and influence of the media in the amplification and attenuation of risk. At worst it could serve merely to aggravate tensions between risk experts and managers and lay publics through its failure to provide a coherent and full understanding of the impact and operations of these plural and symbolic information systems and their relationships with their consumers.

In the case of risk society theory, bold claims are advanced about media influence without empirical substantiation and the media are treated as monolithic (Anderson, 1993, 51; 1997, 188). Research has shown that the news media are greatly differentiated according to their own specific market niches, and are governed by varying economic and political restrictions (Anderson, 1997; Hargreaves et al., 2003). As we explore further in Chapter 4, there are significant variations in nanotechnology coverage across different newspaper types. Different media formats are influenced by different practices and constraints. For example, in the UK risk reporting in the popular press tends to focus more on 'human interest' stories than does the prestige press (Murdock et al., 2003), and television tends to place more emphasis upon items with strong visual appeal (Anderson, 1997).

Whilst individuals do not passively consume media messages, and may negotiate meanings in complex ways, the media may potentially play a significant role in influencing how risk issues become framed (Hornig, 1993; Lupton, 2004; McCallum et al., 1991; Miller and Reilly, 1994; Philo, 1999; Williams et al., 2003). Media reporting of risk may be interpreted in differing ways by individuals, depending upon a variety of factors including their degree of prior knowledge, personal experience and cultural capital. However, there are also problems with over-stating the notion of the 'active' audience. As Philo (1999, 287) maintains,

It would be quite wrong to see audiences as simply absorbing media messages... but it is also wrong to see viewers and readers as effortlessly active, creating their own meanings in each encounter with the text. Our work suggests the media can be a powerful influence on what the audience believes and what is thought to be legitimate or desirable.

Knowledge and perceptions about risk are often influenced by family, friends, colleagues and health professionals and are mediated by factors such as age, gender, ethnicity and social class (see Mythen, 2004; Petts and Niemeyer, 2004; Thirlaway and Heggs, 2005). Moreover, there is great variation in the extent to which individuals rely upon media for particular risk issues. For example, Petts et al. (2001) found that people relied heavily on the media for information pertaining to the millennium bug and train accidents, but not for air pollution issues. People are especially likely to rely on the media as a source of information on the risks of nanotechnologies given that they are invisible.

As discussed further in Chapter 5, recent research on public attitudes towards nanotechnologies suggests that currently relatively little is known about them. To the extent that this is the case, the media are likely to play a formative role in shaping the parameters of public debate. Particularly, when publics have little familiarity or knowledge of an emerging technology, media reporting can provide heuristics for comprehending and assessing the issues (Scheufele and Lewenstein, 2005), although there is no simple, direct correspondence between coverage and attitudes. A number of science reports, including the RS/RAE report, acknowledge the potential for media influence on risk perceptions of nanotechnology; however, scientists' views on the role of the media are often simplistic, based upon the deployment of a 'media effects' model (see Chapter 5). Media (often assumed to be singular) are assumed to have a powerful influence on 'the public' views through 'distortion', 'misrepresentation' or 'amplification' of scientific 'facts'. Yet, as discussed above, the social processes that effect risk interpretation are more complex than can be captured by SARF or the risk society model (Anderson, 2006, 123; Petts et al., 2001, v). Audiences actively engage with and may contest media representations, forming their opinions from diverse sources including, increasingly, the Internet – especially in the case of younger people (Allan, 2006). In media and journalism studies, the emphasis has shifted to the politics of 'framing' news and the examination of the ways in which competing claims makers may seek to impose their definitions of issues, through the systematic presentation of particular claims, facts or values and the downplaying or neglect of others (Allan et al., 2000). The question of how technologies and their risks are represented in the news media during their emergent stage is therefore likely to be crucial for subsequent debate and policy formation.

The findings of our studies present a more complex picture than presented by previous media approaches. Beck's model is too abstract to be easily empirically testable, while SARF suffers from a strong psychometric bias and inadequate problematisation of 'risk'. In the following chapters we indicate that scientists play a more active role in news production than their own representations would suggest. We present a sociologically informed and empirically based analysis of the news media, drawing on new data to illustrate the social relations of news production and the socially mediated character of news portrayals.

Conclusion

In this chapter we have argued that although early nanotechnology coverage shares many similar features to those identified in previous science reporting, it is distinctive in a number of important respects. Given its early stage in the issue attention cycle we have suggested that 'risk' is more open to contestation; currently there is little public knowledge or concern about the issues, and the language and metaphors have yet to be firmly established.

The 'risk society' theory and SARF offer useful starting points in thinking about the media and risk. However, this discussion has highlighted a number of conceptual and methodological limitations with both these models. The foregoing discussion has underscored the need to move beyond a media-centric approach to tease out the complexities and contingencies of social processes and the wider play of social power. This necessitates further empirical research to flesh out, among other things, how news sources seek to influence coverage and the factors affecting their success or failure to secure favourable representation, and a more nuanced account of the impact of media upon public attitudes. Risk is a contested field where news sources are involved in continual definitional struggles. Far from being an 'objective' account of reality, media reporting of risk represents a contentious and negotiated discursive terrain. Accordingly, this requires a multi-faceted analysis of the factors affecting the effectiveness of media strategies over time.

In the following chapters we demonstrate how empirical research into news media coverage of nanotechnologies may begin to address some of these issues. Our studies, supported by relatively small grants, were necessarily limited in scope. As a consequence we were not able to explore audience interpretations or divergences across different media or representations over an extended period of time. Our analysis of source–media interactions was also limited to examining scientists' and journalists' perceptions, since scientists have been shown to be the principal news sources in the formation of the nanotechnology debate within the national daily press. However, we believe that our two studies make a valuable contribution to understanding processes of news production, particularly in relation to print media coverage of nanotechnologies during a period in their rising public salience.

4
News Coverage
of Nanotechnologies

Introduction

Given the novelty of nanoscience and the scale of resulting technologies, media images (news coverage, but also in the entertainment
realm) are likely to be the main way in which most people will relate
to developments in this field. Even when products utilising nanomaterials are directly encountered, their 'nano' qualities are likely to be
only visible as a brand name. The invisibility and definitional ambiguity of 'nanotechnology' has arguably served to constrain media
interest in nano-innovations in recent years. Nevertheless, there are
signs that ongoing debates concerning the scale, novelty and implications of nanotechnologies are beginning to generate greater media
attention (Schummer, 2004; Wood et al., 2007).

Nanotechnology is increasingly being hyped as 'the next big thing',
set to revolutionise our daily lives. However, as mentioned in Chapter
1, despite the growing range of commercial products in the field, its
ethical and social dimensions have been largely neglected in related
debates (Hansson, 2004). Representations of nanotechnology often
draw on imagery (Losch, 2006) which has long been prominent in
science fiction (Thurs, 2007), such as depictions of fantastic voyages
and intrepid miniature explorers navigating the inner workings of
the body. Some disquiet has been expressed that this type of imagery
will come to be associated with the risks of nanotechnology as well
as the benefits, in a fashion similar to that which occurred with
'Frankenstein Foods' being associated with genetic modification (van
Amerom and Ruivenkamp, 2006). Indeed, it is reasonable to suggest

that media coverage of nanotechnologies is already being influenced by the earlier reporting of biotechnologies, which is a legitimate cause for concern.

In this chapter, we explore how nanotechnologies have been represented in the UK national press. More specifically, we examine the range and scope of the coverage, how issues have been framed by different newspapers, the tone of the articles published and the nature of the sources most prevalent in the reporting. As we argued in Chapter 3, we believe it is important to identify the presence – but also the absence – of particular frames and news actors within nanotechnology coverage.

Nanotechnologies in the media

The potential significance of the media is often alluded to by researchers and policymakers who have sought to track the reception of nanotechnologies in society (Bainbridge, 2004; Hett, 2004; Nanologue, 2006; Wood et al., 2003). In the US, the role of media in the reception of these new technologies has been highlighted alongside governmental, industrial and religious influences (Stephens, 2005). A consideration of this role with respect to the ongoing framing of facts, and also values, is unavoidable when examining issues related to 'the public understanding' of nanotechnologies (Cooper and Ebeling, 2007). However, examinations of media content remain limited to a few key studies, undertaken mainly in the US, Canada, UK and the Netherlands (Anderson et al., 2005; Cooper and Ebeling, 2007; Ebeling, 2008; Faber, 2005; Friedman and Egolf, 2005; Gaskell et al., 2005; Gorss and Lewenstein, 2005; Petersen et al., 2008; Schummer, 2004; Stephens, 2005; Te Kulve, 2006; Wilkinson et al., 2007).

In the US context, Gorss and Lewenstein (2005) took a longitudinal approach to news coverage, examining the content of the *New York Times, Washington Post, Wall Street Journal* and *Associated Press* between 1986 and 2004. Their preliminary study found that coverage of nanotechnologies was on the increase, from a few articles in the late 1990s to over 150 in 2003. Furthermore, the study demonstrated that interest had expanded from being restricted to the prestige press which suggests that there is growing public interest in the area. They also found that US coverage tended to be very positive towards the developments. In part, this was due to the increased coverage of new applications and business perspectives but, as they aggregated the tone of the coverage, positively framed stories were very positive whilst negative stories were not especially negative. Though some social, environmental and ethical concerns surfaced in association with various news events, such as the publication of Michael Crichton's *Prey*, these soon dispersed with a return to more positive portrayals. In terms of US coverage, then, nanotechnologies are popular across financial news and business sections, with much less reporting of the attendant ethical, legal and social dimensions (Schummer, 2004).

Few studies have directly compared news coverage internationally. However, those that have suggested that coverage tends to be more positive in the US, whilst risks are slightly more evident in UK reporting (Gaskell et al., 2005). Stephens (2005) analysed 350 randomly selected nano-focused articles from a universe of 1330 articles in a range of US and non-US newspapers spanning from 1988 to mid-July 2004. The majority of newspaper articles sampled (267) were from major US publications, while only 83 were from non-US sources. Unfortunately the author does not provide a breakdown of the source of the non-US articles, though the UK's *Guardian* is referred to as one of the sampled newspapers containing ten or more nanotechnology-focused articles. The preliminary findings suggest that those articles expressing a sentiment towards the development of nanotechnologies tended to see the benefits of nanotechnologies as outweighing the risks. Stephens found that the coverage tended to focus on scientific discoveries or projects in the randomly generated sample, though there were some variations in international coverage with non-US news articles more likely to attend to ethical, legal and social implications (Stephens, 2005).

Friedman and Egolf (2005) focused specifically on newspaper coverage of the social, health and environmental risks associated with nanotechnology in the US and the UK between 2000 and 2004; this numbered 71 articles in the former and 50 in the latter. Though this is a small percentage of the total coverage relating to nanotechnologies, it is noted that the number of articles exploring these risks was increasing between 2003 and 2004. Interestingly when the actual content of the articles was examined in this study, it was determined that headlines were frequently providing the most negative overtones (Friedman and Egolf, 2005). The preliminary evidence, then, suggests that UK and US coverage has been relatively similar, though UK coverage has tended to be slightly more negative, and with slightly more reference to potential social implications (Friedman and Egolf, 2005; Gaskell et al., 2005).

Studies thus far have been largely limited to US and, to a lesser degree, UK news coverage. Exceptions include Te Kulve's (2006) study of Dutch newspaper reporting of nanotechnologies, which found stories evolving from a scientific emphasis to broader social and economic implications, alongside risks. As with other European studies, the coverage was found to be significantly more negative in tone compared to US reporting. Across broader media genres, studies have analysed magazines (Schummer, 2004), popular science books (Schummer, 2005) and science fiction (Miksanek, 2001; Schummer, 2004; Thurs, 2007) or taken a more focused approach by examining, for example, the images used to accompany articles (Losch, 2006). It is notable that the tendency has been to focus on forms of print media, rather than television, radio or the Internet.

In this regard, it is worth observing that a weakness of some prior studies on science reporting has been the focus on prestige or elite coverage, namely because it tends to be seen as 'agenda-setting' and therefore politically influential (Carvalho, 2007). Largely neglected has been the news reporting of the more popular 'tabloid' or 'red-top' newspapers at the other end of the readership spectrum, despite their much higher (on average) circulation figures. Prior studies have rarely focused on how nanotechnologies have been portrayed across a range of mainstream print media and this is a significant oversight where the development of public attitudes to nanotechnologies is concerned, since newspapers vary significantly by readership. In the UK, for example, 'red-top' and mid-market coverage is likely to reach not only the largest number of readers, but also those who may have fewer years in formal education in their personal backgrounds than would be typical for readers of the so-called 'prestige' titles. As discussed in Chapter 3, it is important to differentiate between different news media according to market, economic and political stratification, especially where controversial science is concerned. Mid-market newspapers like the *Daily Mail* played a significant campaigning role in generating controversy around GM products, for example (Cook et al., 2004).

Longitudinal quantitative-based studies, although essential, may not always reveal the finer nuances of coverage. Factors such as how sources or documents are used, where coverage is appearing within individual newspapers and how it differs (or is influenced) by the type of correspondent reporting on it are important, as the news media play a powerful political role where science is concerned. These factors have underlying implications for how funding is allocated, regulation is developed and issues highlighted for public attention (Carvalho, 2007; Nisbet and Lewenstein, 2002). Expanding analysis to include diverse mainstream media sheds light on the potential implications in terms of style of reporting. Elite or prestige newspapers frequently have dedicated science and technology sections, as well as specialist editors and correspondents. The question of how the coverage appears in these newspapers compared to that in the popular press warrants investigation (Nelkin, 1987).

Although investigations into the role of the journalist in science news have an extensive history, these have not extended into the field of nanotechnologies. This is a significant oversight when much of the policy-making in the field encourages scientists to engage with the media in order to reach publics (Petersen et al., 2008). It has also long been recognised that where scientific coverage is concerned there is a shared culture, 'shaped by the cooperation and collaboration of several communities, each operating in terms of its own needs, motivations and constraints' (Allan, 2002, 2008; Carvalho and Burgess, 2005; Nelkin, 1987, 11; Nisbet and Lewenstein, 2002). That said, journalists can be a difficult group to engage with from a research perspective. Their professional constraints may mean that interviews are hard to schedule. Specialist correspondents are relatively low in number, may be selective about their participation in research and wary of social scientists following the wide-scale academic interest and controversy recent science coverage has generated, such as that surrounding biotechnology. In the case of nanotechnologies, their relative novelty may mean that general correspondents who report on the issue as a news item may fear they lack the relevant background knowledge or context to contribute effectively. Thus far Cooper and Ebeling's (2007) study is the only other study apart from ours to have included interviews with journalists regarding nanotechnology coverage. They interviewed eight UK-based journalists who had authored articles on nanotechnology and worked in financial and science journalism on newspapers, magazines, periodicals or newsletters.

Having briefly surveyed the literature in this field, we now turn to comment on the background to our study which involved a content analysis of news coverage. In the UK, 2003 to 2004 proved to be a formative time period in which to examine how and why nanotechnologies were being covered in the news media. Until then, in the UK and internationally, there had not yet been a public outcry about these technologies and, as we shall see in the next chapter, public awareness remained low. However, we began to witness increasing media interest, a focus on the implications of nanotechnologies from learned societies and interest groups, and the intervention of a key UK public figure, HRH Prince Charles, whose comments on nanotechnologies were evidently deemed to have considerable 'news value' by the British press. Prince Charles had taken on a similar interventionist role in relation to science coverage in 1999 when he published a news article questioning GM crops in the *Daily Mail* (Adam, 2000b). We may also draw similarities to the 'episodic' coverage afforded to biotechnology up to the 1990s, identified by Nisbet and Lewenstein (2002), which contributed to later media debates. Thus we identified a setting where a media agenda for nanotechnologies was being constructed and an opportunity to examine the manner in which media coverage was developing on the subject. It also provided some indicators of how the initial parameters of public debate were gradually beginning to consolidate.

Nanotechnologies and news production

The content analysis of news coverage undertaken in our project occurred during a 15-month period, between 1st April 2003 and 30th June 2004. This time period was selected since a number of nanotechnology-related issues began to arise in the press, including high-profile comments made by HRH Prince Charles, the commissioning of the joint RS/RAE study focusing on nanotechnologies, and the release of a number of fictional books and films in the area. This provided impetus for news coverage of the topic, which has not been sustained to the same level since. Of particular value here is the way in which it reflects on news reporting of the issue during its 'formative' period (Anderson et al., 2005), a time when many members of the public in the UK were being introduced to the term 'nanotechnology'. Our study remains one of the few that has focused specifically on news media representations. As mentioned, research has examined coverage both within the UK (see Gaskell et al., 2005) and at an international level (see Faber, 2005; Friedman and Egolf, 2005; Gorss and Lewenstein, 2005; Stephens, 2004, 2005) but these studies have not assessed such a broad range of UK-based newspapers nor gone on to interview journalists/editors themselves.

The newspaper articles were selected for the sample on the basis of the inclusion of four keywords: 'nano', 'nanotechnology', 'grey goo' and 'nanobot/nanorobot'. These had been established as appropriate for identifying articles following a pilot search using three newspapers. The final sample included ten UK-based national daily newspapers and eight UK-based national Sunday newspapers, encompassing 'red-top', 'mid-market' and 'opinion leading' publications. The daily newspapers sampled were *The Times, The Guardian, The Daily Telegraph, The Independent, The Financial Times, The Daily Mail, The Daily Express, The Daily Mirror, The Sun* and *The Daily Star*. The Sunday newspapers sampled were *The Sunday Times, The Observer, The Sunday Telegraph, The Independent on Sunday, The Mail on Sunday, The Sunday Express, The Sunday Mirror* and *The News of the World*. The articles to be analysed were identified via LexisNexis Professional and NewsBank Newspapers – UK to ensure that the sample was comprehensive and complete.

In total, 344 newspaper articles were identified across the sampling period. Each article was analysed using a coding schedule. The coding schedule recorded quantitative details including newspaper title, date, page number, author attribution and sources cited or referred to. It also examined more qualitative aspects of the data, including the leading news 'frame', the 'tone' of the item, and attribution of 'risks' and 'benefits' associated with nanotechnology. These aspects of the schedule were based on a similar format developed by Stephens (2004) for his analysis of news media frames, but also included categories specific to the UK context. As the more qualitative aspects were vulnerable to subjective interpretation on the part of the coder, a sample of 12 per cent of the newspaper articles was coded separately to ensure inter-coder reliability. We took hard copies of each newspaper article to provide a 'feel' of the original article, which can sometimes be neglected using electronic versions alone (Hansen et al., 1998). The analysis was performed using an N5 database of the articles and univariate descriptive analysis bivariate analysis using SPSS11.5.

Level of coverage

Perhaps not surprisingly in the light of the discussion above, the press coverage during the period under scrutiny was concentrated in a relatively small number of 'elite' newspapers. Eighty-six per cent of articles ($n = 296$) originated from the ten sampled daily newspapers and 14 per cent ($n = 48$) from the eight sampled Sunday newspapers. Of the daily newspapers, the vast majority of articles ($n = 255$, or 74 per cent) appeared in the elite press, which have relatively low circulation figures, while the rest ($n = 89$, or 26 per cent) appeared in the more popular (i.e. mostly high circulation) newspapers. *The Guardian* featured the largest number of articles ($n = 81$, or 24 per cent) of the daily newspaper coverage. This was followed by *The Times* ($n = 65$, or 19 per cent), *The Financial Times* ($n = 47$, or 14 per cent), *The Independent* ($n = 36$, or 10 per cent) and *The Daily Telegraph* ($n = 26$, or 7 per cent). However, references to nanoscience and nanotechnology appeared in all of the sampled newspapers, with the exception of *The News of the World*, the highest circulating Sunday red-top newspaper, which did not feature any articles containing the keywords during this sampling period.

This concentration of coverage to a few elite daily newspapers with relatively small distribution figures means that the visibility of the issue had been largely restricted to the relatively small middle-class and business groups who can be assumed to be the readers of these newspapers. The majority of readers of *The Guardian* (61 per cent) and *The Times* (63 per cent), for example, are from a professional or skilled group (Newspaper Marketing Agency, 2005).

Type and tone of coverage

As mentioned previously, the articles were examined in terms of the news media frame presented in the article. As discussed in Chapter 2, framing aspects of news coverage is a significant process of negotiation and mediation within the journalistic community as they seek to channel information into an appropriate and newsworthy media format. A range of categories emerged as is illustrated in Table 4.1.

We can see that, apart from the interest focusing on Prince Charles' intervention, a science-related frame dominated much of the coverage over the period, indicating a strong news interest in the scientific implications of nanotechnologies. There was also considerable attention devoted to nanotechnology applications, with medical discoveries and projects featuring as a focus. However, the comparable dominance of the 'science fiction and popular culture' frame and the 'scientific discovery or project' frame sees science competing with fictionalised notions of the developing field, reflecting uncertainty about where the boundaries between the fictional and the factual aspects of nanotechnologies lie (see Anderson et al., 2005). 'Celebratory' articles focusing on awards and the like for the most recent work in the field, alongside 'business story' and 'funding' articles, demonstrate a clear interest in the economic implications of nanotechnologies (see also Cooper and Ebeling, 2007). Furthermore the presence of nanotechnologies in articles related to possible careers and education, and even the latest children's science fiction film at the cinema, reflects the hyping of nanotechnologies for younger readers. This encompasses career prospects for A-level science students, for example, as well as the latest fictional mini-beast (see also Berube, 2006).

Interestingly, however, there were significant variations in news frames across different newspapers. *The Guardian* coverage tended to focus on a 'scientific discovery' frame ($n = 17$, or 31 per cent) and a 'social implications' frame ($n = 15$, or 47 per cent) tracking the emerging developments and their consequences. *The Financial Times*, predictably, had the largest single number of articles with a 'business story' frame ($n = 20$, or 38 per cent) and a 'funding' frame ($n = 10$, or 37 per cent). Whilst we may assume that the 'science fiction and popular culture' frame would have appeared in mid-market and red-top coverage, it in fact appeared mainly in *The Times*, *The Independent* and *The Guardian*, accounting for 69 per cent of all coverage, while stories with an 'educational' frame appeared mostly in *The Guardian* ($n = 12$, or 46 per cent). Coverage in the red-tops varied, with the three articles taken from *The Sun* all dominated by the 'scientific discovery or project' frame, whilst *The Mirror* displayed a mixture of coverage with its 18 articles falling across seven of the ten frames. In other

Table 4.1 News frame by three-month quartile

News Frame	April to June 2003	July to September 2003	October to December 2003	January to March 2004	April to June 2004	Total (Percentage)
Science Fiction and Popular Culture	19	12	8	10	6	55 (16)
Scientific Discovery or Project	12	11	9	12	10	54 (16)
Business Story	8	6	9	18	11	52 (15)
Prince Charles' Interest	29	4	3	1	2	39 (11)
Other	7	7	7	10	2	33 (9.5)
Social Implications and Risks	20	3	1	5	3	32 (9)
Funding of Nano	7	4	4	7	5	27 (8)
Educational or Career Advice	3	5	7	7	4	26 (7.5)
Medical Discovery or Project	4	7	3	1	5	20 (6)
Celebratory	1	0	2	0	3	6 (2)
Total (Percentage)	110 (32)	60 (17)	52 (15)	71 (21)	51 (15)	344 (100)

words, readers are offered varying introductions to nanotechnologies according to the newspaper they read, with some newspapers presenting coverage of scientific projects alone and/or their potential implications, while others have a multitude of frames in action from the 'fictional' to the 'factual'.

We also sought to assess whether there were differences in the types of frames appearing across time. To do this we examined the data across three-monthly periods, that is quartiles. The largest number of articles appeared during the first quartile ($n = 110$, or 32 per cent), from April to June 2003 and, although there was a decline in levels of later coverage, it remained relatively constant throughout the remaining periods, at an average of 58 articles per three-month period. Twenty-nine (i.e. 74 per cent) articles framed by 'Prince Charles interest' appeared, somewhat predictably, between April and June 2003. The intervention of this public figure appears central to the rising prominence of nanotechnologies on the media agenda at this time, despite some scepticism and criticism in the reporting of both his intervention and apparent knowledge (see Anderson et al., 2005). His role is similar to that taken in the 1980s by the then Prime Minister Margaret Thatcher in the initial framings of UK climate change coverage, where the issue shifted from scientific to, in this case, a political news agenda (Carvalho and Burgess, 2005). Prince Charles was depicted in some news stories as an interfering and ignorant public figure, in others as a proactive, politically aware individual challenging and questioning the political support at that time accorded to the field (see Anderson et al., 2005). The impact of his role on the nanotechnology news agenda illustrates how a celebrity figure may shape the news agenda and how, despite scepticism and criticism within media coverage, it may draw attention to nanotechnologies across different sections of the press.

Perhaps reinforcing this point, a large proportion of the articles with 'social implications and risks' and 'science fiction and popular culture' frames also appeared during the first quartile period (20 of 32 or 62 per cent and 19 of 55 or 34 per cent, respectively) when 'grey goo' style notions initially dominated coverage regardless of specific attention to Prince Charles' initial intervention. It is worth noting that 'grey goo' was not a phrase used by Prince Charles, though his dystopian concerns quickly became associated with grey goo scenarios – that is, the perception that small self-replication machines could quickly multiply and become destructive. Both the frames 'Prince Charles's interest' and 'social impacts and risks' received little coverage during the rest of the sample period. It is also worth noting the increase in the number of articles with a business focus. Coverage of scientific and medical discoveries or projects remained at a relatively constant level throughout the duration of the sample, while 'business story' peaked in January to March 2004 covering a quarter of that period's reporting of nanotechnology issues ($n = 18$, or 25 per cent). It is notable that news stories did not belong to a specific news genre, but rather were positioned across a range of genres – including the hard news, science, health, educational and business sections.

When the articles were coded for their attribution of risks and benefits there was found to be a strong sense of optimism regarding the potential of nanotechnologies. The majority of articles ($n = 132$, or 38 per cent) were coded as 'benefits outweigh risks'. In these articles, although potential risks were examined, the focus was on the potential offerings of nanotechnologies. A much smaller number of articles ($n = 38$, or 11 per cent) depicted the risks as outweighing the benefits, with a slightly larger proportion of articles ($n = 56$, or 16 per cent) arguing that the risks and benefits needed consideration but were still unclear. Finally a small portion of the coverage ($n = 16$, or 5 per cent) recognised that progress in the field was limited by technological factors rather than ethical, legal or social factors. Overall then, despite a number of articles considering the risks, uncertainties and limitations of nanotechnologies, the general picture painted by UK coverage was largely optimistic.

Sources and authors

As the literature on science communication reveals, much can be learnt about the framing of information on science and technology issues by examining both the authorship of stories and the nature of the sources that are consulted. Our study revealed that general correspondents most often authored articles, with only 13 per cent ($n = 43$) of the sample written by science correspondents or editors and just 4 per cent ($n = 13$) by technology correspondents or editors. News items in the 'elite' press (e.g. *The Guardian*) were more likely to be authored by a science correspondent, while news items in the popular press were more likely to be written by a political or non-specialist news reporter. One would anticipate that journalists with a specialist background would be better positioned to present a more scientifically nuanced account of nanotechnologies with attention to benefits and risks. While we did not explore in detail the quality of reporting between different news media, we noted that the 'red top' and popular media tended to provide less scientific detail and focused more upon extreme scenarios.

Prince Charles' intervention appears to have aided the shift in the reporting of nanotechnologies from specialist to general or political correspondents. Although it can be argued that journalists from non-specialist or non-science backgrounds are capable of reporting on the issues in an appropriate way (Nelkin, 1987), they are more likely than specialist reporters to be unfamiliar with the novel questions raised by this field. It may also be argued that this may predispose these journalists to compare nanotechnologies to earlier technologies with which they are familiar. That is to say, certain ready-made frames help to highlight ways to render social implications and risks newsworthy.

Table 4.2 illustrates some of the headlines that appeared in the UK national daily press at the time. The single most prevalent documentary source referred to in our sample period was Michael Crichton's novel *Prey*, with 22 references. Given that certain metaphors and images in news reporting can sometimes blur the distinction between science fiction and science fact (Petersen et al., 2005), the salience of references to this novel and its associated imagery of deadly nano-swarms is striking. The RS/RAE study was the next most prominent

Table 4.2 Examples of headlines from UK national daily newspapers

'Grey goo science is used in sunscreen', *The Mirror*, May 26th 2003, p. 22.
'The real goo', *The Times*, June 24th 2003, p. 17.
'Brave New World or Miniature Menace: Why Charles fears grey goo nightmare: Royal Society asked to look at risks of nanotechnology', *The Guardian*, April 29th 2003, p. 3.
'Don't be afraid of the grey goo', *The Financial Times*, April 30th 2003.
'Who's afraid of a miniaturised Racquel Welch?', *The Independent*, April 29th 2003, p. 17.
'Spare us all from Royal nanoangst', *The Telegraph*, April 30th 2003, p. 20.
'Charles Caught in Grey Goo Row', *Daily Mail*, June 12th 2003, p. 32.
'Exclusive: Labour Donor tells Prince to keep views to himself: keep your nose out Charles', *Express*, June 16th 2003, p. 2.

documentary source, cited by 16 of the newspaper articles (RS/RAE, 2004).

Where sources were utilised in articles, 177 of the articles referred to, quoted or cited an individual source. They were dominated by scientists and researchers ($n = 173$) working in the field, as found in previous studies of science reporting (see Chapter 3). These sources tended to have a high prominence in articles, often being the first or second source referred to an article, as opposed to those trailing in later sentences or paragraphs. Of the 177 articles which used such sources the majority offered direct quotes, with 172 of the articles using a comment from the source within the article. The use of quotes, particularly from scientists and researchers, lends credibility to articles by suggesting that information is unmediated; that is, direct from those 'in the know'. Interestingly, however, scientists frequently complain about the quality of news reporting, which is seen as negative, biased, 'sensational' and as 'distorting' or 'hyping' science fact, which downplays their own considerable role in framing issues (see Chapter 5).

In terms of individual personalities that featured prominently in the coverage, aside from Prince Charles, few characters dominated the coverage. Scientists, in particular, often featured only once or twice before disappearing from news reports. The most frequently cited source was Lord Sainsbury ($n = 16$), the UK Science Minister at the time, Ian Gibson ($n = 8$) MP, Eric Drexler ($n = 6$) (author of *Engines of Creation: The Coming Era of Nanotechnology*) and Pat Mooney ($n = 6$) (Executive Director of the ETC group). The scientists that featured most prominently included Sir Harry Kroto ($n = 5$), Professor Mark Welland ($n = 5$) and Baroness Susan Greenfield ($n = 4$) but as these numbers indicate, no single scientist especially dominated the coverage during the time period sampled. Politicians and governmental representatives featured prominently, while spokespersons of groups like Greenpeace and the Royal Society were relatively unrepresented. Although such organisations have become increasingly able where media relations are concerned (Nisbet and Lewenstein, 2002), our data confirm that in the early stages of issue development a more exclusive range of news actors appear to be influential (Nisbet et al., 2003).

The comparative emergence of a science fiction/popular culture framing was related perhaps to the 'event-driven' nature of the coverage (Gorss and Lewenstein, 2005) whereby the announcements of personalities and prolific instances of popular culture, such as the publication of *Prey*, attracted attention to nanotechnologies. Here, then, a science issue had broader news value and perhaps the lower number of articles authored by science and technology correspondents becomes more explicable, as does the broad range of sources utilised from non-science backgrounds.

Journalists' perceptions of nanotechnologies

In conjunction with the content analysis of news coverage, we conducted in-depth interviews with five journalists who had all contributed to articles on the topic of nanoscience and nanotechnologies, in order to ascertain their views on media coverage and public perceptions of nanotechnologies. Following university ethical clearance, the journalists were contacted via email with an accompanying letter explaining the project and requesting their participation. We then scheduled an interview with them, either in person or via telephone. That we were able to undertake only a limited number of interviews with journalists reflected the professional constraints under which journalists operate. Our interviews and emails were at times neglected, interrupted or cancelled. The poor response rate also reflects the relatively high turnover of journalistic staff working within media outlets and the small pool of individuals working in the field, noticeable in the relatively small number of attributed authors in the original media analysis. Some general correspondents indicated that the news reports on nanotechnologies that they authored were beyond recollection. At the time of the interviews nanotechnologies were not generating a large amount of coverage and this contributed to some journalists claiming unfamiliarity with the issues and being disinclined to participate. However, the five journalists who did participate were all prominent UK-based science and environmental correspondents, working on national print publications and newspapers – including an individual who had played a key role in breaking the initial headline coverage. Their views are therefore insightful in relation to the debate on nanotechnologies. In the following paragraphs, extracts from our data have been kept anonymous in order to protect the confidentiality of participants.

Asked what spurred their interest and coverage of nanotechnologies, the responses echoed the general trend that had been indicated in press coverage. First, news hooks had become apparent as new technologies emerged, be they fictional or 'celebrity' based and, secondly, the governmental and scientific establishments' focus on the area had provoked interest and, at times, an air of suspicion. This is alluded to in the first extract:

> Nearly all the articles I've written on nanotech were a response to something that happened, e.g. Cornell producing the micro guitar. Drexler was a stimulus for a lot of articles because he said things and then people would react and royalty got involved. The tenor of the coverage went from 'gee whizz' to quizzical (well I don't know....) to slightly distrustful. A lot of stuff was in response to the Royal Society working party being set up. Lord Sainsbury and Ann Dowling had a press conference for journalists. I asked them what was different about nanotech and they said they weren't sure. They didn't seem to think it was that different. But there must be something truly novel about it. The investment went from practically nothing to billions of pounds so something must be going on. Lots of things are going to happen.
> (Science Editor/Correspondent A, UK-based Quality Newspaper)

> The initial fears [were] probably mostly stirred by Crichton's book, nanoreplicators. Given that Mother Nature can't make nanoreplicators except for sort of very simple systems, self-assembling molecules and so on but nothing very....a lot of the controversy at the start was basically the science establishment really trying to sort of really argue for a bit of a rational debate and I think now after the...I think the Royal Society/Royal Academy exercise was quite a good one. They probably wasted 95% of their time, because I think the only interesting bit that came out was the particulates question. But I think it did a lot to sort of get the debate out in the open and...so it was unusual to report something like that which was quite a good synthesis, I think, of pressure groups versus establishment and identifying areas that needed more research.
> (Science Journalist B, UK-Quality Newspaper)

Both of these journalists, then, were able to highlight specific incidences that precipitated coverage, as well as indicating a general sense of an emerging news story, as governmental, fictional and financial interest converged. The journalist in the first extract described news coverage as 'a response' to that occurring in the field, while in the second extract the issue of nanoparticles ('particulates question') is also raised. This was the only concern that a number of interviewees appeared to take seriously, often raising it themselves during the interviews. Interestingly in our newspaper analysis, comparison to other particle-related risks, such as asbestos, was not a prominent theme within the coverage. That said, it was raised by both journalists and experts throughout our interviews as one of the potential risks of the emerging developments (Wilkinson et al., 2007).

Nanotechnologies present an interesting case in the evaluation of risk. In Chapter 3, we outlined the 'rationalist' position to notions of risk, whereby the 'riskiness' of technologies may be measured against the collection and analysis of appropriate data (Hornig, 1993). In the case of nanotechnologies, this rationalist argument is lurking in the media headlines. Scientific sources continue to defend the technologies with the argument that despite current limitations in knowledge, credible research and evidence about the nature of attendant risks is forthcoming. In other words, although a clear rationalist, evidenced argument cannot be made at present, due to the emerging and developing nature of the technologies, the scientists tend to be convinced that such will be the case at some stage in the future. Technologies deemed to be revolutionary are much more likely to secure attention but, at the same time, scientists evidently recognise the need to offer reassurance that a body of accurate, factual knowledge is being accrued that will prove that risks posed by nanotechnologies are no greater than those of other areas of technological innovation. The implication of this is that scientists can make very few credible claims of 'wrong thinking' within the media in relation to the risks associated with nanotechnologies (Hornig, 1993). The so-called 'grey goo' scenario is one of the few portrayals that scientists appear confident in actively rejecting.

In this sense, discourses around nanotechnologies tend to be indicative of the 'subjectivist' position towards risks (Hornig, 1993), whereby the emphasis is on the socially constructed nature of risk itself. Journalists, as well as the scientists, are actively grappling with the problem of representation. One study that has explored some of the metaphorical imagery surrounding nanotechnologies suggests that images mirror concerns that particle-related risks could become an increasing news focus. 'Nanoparticles as the new asbestos' has been a slogan of sorts for a variety of stakeholders, highlighting explicit policy and research implications (van Amerom and Ruivenkamp, 2006). In contrast, 'grey goo' imagery, as mentioned, has been less influential amongst communities of expertise (van Amerom and Ruivenkamp, 2006), a finding that is supported by our interviews with journalists and scientists alike. There are concerns that the spotlight on 'grey goo' and self-replication arguments has diverted attention from the more likely scenarios, such as surveillance and human enhancement and particle-related risks (Wood et al., 2007).

Whilst the 'grey goo' concept owes its lineage to a combination of fictional and scientific prophecy, it is increasingly recognised as unrealistic (van Amerom and Ruivenkamp, 2006). An environmental correspondent described a cycle of coverage, again stressing the political emphasis on nanotechnology supported by a prominent news hook associated with the intervention of Prince Charles. This cycle of coverage, he believed, had 'broken' by the time of the interview:

> It [the press] tends to report what everybody else is reporting, what the politicians are interested in, and the politicians are interested in what the press is interested in and there are a lot of self-reinforcing circles and it's quite hard to break into those circles... it can be an intervention, like Prince Charles' intervention when he wrote for the *Independent on Sunday*. That broke the cycle and brought it up a bit, but it didn't really take off... so, um, there's two things that have to happen. One, something has to happen to break into the public domain in a big way and, secondly, it has to be kept going by constant stories.
>
> (Environmental Correspondent C, UK-Quality Newspaper)

Thus for these journalists attention to nanotechnology issues appeared to have been drawn to by stakeholders, interventions by members of the scientific establishment, a member of the Royal family and portrayals of fictional futures. Gorss and Lewenstein (2005, 22) describe a heightening media interest in the US following publication and release of fictional accounts of nanotechnologies, the intervention of a key scientist and an agenda-setting piece in the *Washington Post* which drew attention to environmental issues, suggesting coverage is 'event-driven' rather than 'issue-driven'. In fact, when asked whether nanotechnologies raised any new regulatory, social or ethical issues, the correspondents stated that these were limited and/or familiar, particulate safety being the only area mentioned as warranting further investigation. Journalists may then operate with some degree of professional insecurity if they do not follow 'the pack' where a science story is concerned. For these interviewees, nanotechnologies may or may not be a particularly novel or attractive story, but something akin to a pack-mentality ensures that they will monitor developments nonetheless (Cooper and Ebeling, 2007).

Journalists, when asked about the types of sources they had used or would be likely to use, indicated that they saw reputable scientists as key sources. At the same time, however, they were more sceptical about the value of consulting stakeholders or pressure groups. For this correspondent, the utilisation of appropriate sources raised a further issue, which he discussed at length throughout the interview:

> You know, if you say to me oh let's get a nanotechnologist, what does that mean? Does that mean a molecular biologist? Does that mean a guy who, you know, is doing atom force mycotrophy? Does that mean, you know, someone trying to squeeze more features on microchips? Does it mean a solid state chemist looking at quantum effects in big particles? You know, there isn't such a thing as a nanotechnologist, you know, as it encompasses so many different fields of science.
>
> (Science Journalist B, UK-Quality Newspaper)

Throughout this interview the correspondent rejected the term 'nanotechnologies', criticising it in a fashion similar to many scientists (see Chapter 5). This concern around definitional problems, reservations, scepticism and motivations around the term 'nano' has also been expressed by journalists interviewed elsewhere (Cooper and Ebeling, 2007; Selin, 2007). In the above case, however, the interviewee identified the term 'nano' as a cynical rebranding exercise. In his words:

> I'd say that my only problem with nanotechnology is just it's the world's most stupid word. I mean if I said to you that bullets, pennies, bees, marbles, Tintacs were centi technology would that mean anything? Without re-branding these things that way it

wouldn't mean a darn thing would it? So I...I think it's sort of saying in a vague way gee whiz we can do things at the nano level aren't we clever but it's not really any more...I don't think it actually is a label that particularly helps the public understand what's going on...You know the whole body.... Everything in the body is exotic nanotechnology in the cell anyway so, you know, it actually also falsely conveys this impression that it's sort of an artificial...

Interviewer: Why do you think all this re-branding has taken place?

Well it's happened to the poor old chemists several times with molecular Biology and solid state Physics. It's just a parlour game played, you know, in the great effort to get science funding really. It sounds a hell of a lot better if you call it nanotechnology than if you called it applied colloidal science, which is what a lot of things used to be called 20 years ago.

(Science Journalist B, UK-Quality Newspaper)

These comments reveal the constructed nature of 'nanotechnology',whereby what is defined as 'nano' – whether dictated by funding or publicity – is subject to negotiation. In the case of this evolving and contested field, bright scientists may well be able to trademark innovation within their work, whilst isolating any risks incurred through traditional scientific practice. In other words, they may well be able to operate a balancing act whereby they hype the novelty of the field for future investment, whilst at the same time equivocating the risks of nanotechnologies with other well-regulated, long-term activities or practices. Elsewhere we have reflected on the manner in which some risks are compared by scientists to prior health and safety issues, be it former experiences of 'risky' substances, such as asbestos, or to the everyday health and safety decisions made in laboratory settings (Wilkinson et al., 2007).

When the journalists were asked how they sought to explain nanotechnologies to their readers, a number of methods were discussed. These included the use of particular metaphors, similes or images, all of which were identified as having their own particular strengths and weaknesses:

I mean it's quite hard to convey the scales we're talking about and the good old human hair's often invoked to help people get a grip on how small is nano. But the problem with metaphors is that they can often, you know, they give you a limited flavour and then once you start to look a bit more deeply you find the metaphor falls apart badly. Particularly quantum effects, I mean you know they're almost impossible to explain and you just have to be... you just have to say this is the way it works and if you want to understand it, tough!

(Science Journalist B, UK-Quality Newspaper)

Another explained,

I use metaphor or simile all the time. They're tools and we'd die without them. That's the whole point of explaining – you do it by metaphor, analogy and anecdote. Richard Feynman first raised the question of how small you can do it. He gave the example of putting the Bible on the head of a pin. We can't imagine an atom but we can imagine the Bible being very small. However, there is a danger when metaphor determines how you think and how you react. Then it can imprison your understanding e.g. Dawkins and the selfish gene. There is no other way of explaining something that you can't see. If we can't see it microscopically we don't care. It's easy to report things people can see or imagine.

(Science Editor/Correspondent A, UK-Quality Newspaper)

Although our content analysis did not directly analyse the images associated with the various newspaper articles analysed, journalists stressed the importance of 'fictionalised' imagery for interventions that are invisible to the human eye. This has been highlighted previously. Losch (2006, 393), for example, stresses that nanotechnology remains a 'highly visionary topic', with a variety of publications reliant on the same or similar imagery despite their lack of realism as an explanatory tool. He suggests that these common images act as mediators between scientific, economic and media discourses, creating a 'communicative space' to articulate alternative conceptions of the future of nanotechnologies (Losch, 2006, 394). In a similar manner, Thurs suggests that scientists, journalists, educators and others have used science-fiction images to render the science underpinning nanotechnologies 'noteworthy, interesting and important in modern culture' (2007, 66). This was supported by some of the comments journalists made:

> It is impossible for newspapers not to get carried away by the submarine going into blood cells image. One non-science colleague saw an image and thought it was a photograph but it was a piece of artwork. The tell-tale pictures of cranes and machines fascinate people. The very crude level, this could run away with us, alarm, the grey goo that lasted about 8 hours. We all wrote pieces that this was unlikely to happen. It didn't run for very long... Nano is referring to invisible things – actions, hardware, renewables. With things like diabetes people can relate to it more as everyone has some direct experience of it through people they know and so they are prepared to make more effort to understand it at the microscopic level. If you say to people how does aspirin treat headaches they don't want to know. It's like a black box.
>
> (Science Editor/Correspondent A, UK-Quality Newspaper)

> I mean the key thing is that you've got to just explain what it's about. It's about the extremely small scale that we're dealing with here. I don't think... well I don't think the human brain, actually, is very good at grasping extremely small or extremely big things. And so in that sense it is difficult, it can be difficult, to explain. But I think most people can grasp it. The idea that you are sort of building things from the bottom up rather trying to trickle them out from the top down. I think it's... the idea is you've got to start sort of explaining what uses things are going to have, make it relevant to people so they start to understand it.
>
> (Science Editor/Correspondent D, UK-Quality Newspaper)

As the interviews progressed, the journalists discussed the common techniques they used to explain the complexities of nanotechnologies to their readers. We asked them to explain their role in engaging publics with science and technology. According to some scientific commentators, journalists should promote the public understanding of science. However, our respondents tended to see their roles somewhat differently, as is evident in the comments of this journalist:

> I think it's important to get things right. Equally I don't think anybody should ever take the media for granted as explainers of

science as it were. I think the scientists have to explain things to the media to make sure that we get it right. I mean at the end of the day the media are trying to sell newspapers or win viewers for television programmes or whatever and we're going to do things that we think are new and interesting as a result of that. We're not in the business of sort of public communication of science. There's a big difference between journalism reporting and public information as it were.

(Science Editor/Correspondent D, UK-Quality Newspaper)

Other journalists expressed similar views:

Journalists don't really position themselves in the public engagement debate. They've got the rather parochial outlook of selling newspapers, or getting people to turn on the television, or listen to the radio. You know it's not really about ... there is no mission to enlighten or reveal the truth! I'd love to think so, but it gives a distorted picture.

(Science Journalist B, UK-Quality Newspaper)

The press is an unconscious agent of democracy. It should be free to do what it likes – free to be wrong and to be evil. The press has a role in supporting democracy. The idea that the press has a responsibility to science is dodgier. The press has a responsibility to the citizen ... We have a very large responsibility but it's not my responsibility to explain science to the public, it's to get them to read the stories. When I'm talking to scientists and they ask me to show them the article before it's published to check for accuracy I tell them it's my article and I'm responsible for it. I won't show them before it's published and if they don't like it then they won't talk to me again. I make sure I understand what they say (and what they mean to say) otherwise I don't write it. When it comes to it, is this good or bad? It's up to the readers to make their own minds up.

(Science Editor/Correspondent A, UK-Quality Newspaper)

Conveying the emerging field was a point of contention for journalists, both in terms of the relative appropriateness of the term 'nano' and the means of how to explain its significance most effectively. What is clear is that for these journalists, at least, the role of 'explainer' was not one they approve of necessarily. This stands in contrast to some assumptions made within parts of the scientific community, and points to the need for a changing conceptualisation of the journalistic role in science communication.

As we have seen, at the outset of debates around nanotechnologies, journalists encountered serious definitional challenges. In addition to non-science stakeholders, two powerful science groups, namely the RS/RAE, began to take an active role in defining the pertinent issues. The commissioning of the joint RS/RAE (2004) study saw a range of groups come together in an attempt to navigate through some of the more controversial issues. For science and environmental correspondents there were clear comparisons to previous high-profile controversies, such as genetic modification and BSE. It appeared that for some this had meant notable changes in the actions of news sources with whom they were dealing.

> I think that because nano was confronted early on by the scientific establishment (the Royal Commission and the citizens' panels) they tended to diffuse anxiety. Anxiety is roughly disproportionate to the amount of information that is available. Something like 10 years ago government scientists were denying BSE and the public got more anxious because there was no obvious evidence. The public are not so dumb as some would have them. GM was a different case, a case where the press had been saying what can this do to plants and no-one batted an eyelid and then the situation exploded. The nanotech show was one where the scientists came out fighting. The scientists were willing to talk. And they gave sensible answers addressing the fairyland aspect of a lot of science.
> (Science Editor/Correspondent A, UK-Quality Newspaper)

> In fact in nanotechnology there are some pretty good people around who I think have cottoned on to the fact that ah ... I think they've learned from things like the GM experience on that and have actually realised that they need to sell their work at an early stage because it does have the potential to scare people. And somebody like say [name removed] has been extremely good at engaging with the media, explaining things, and ah making it

clear from the beginning. There's also at [name removed] who does a lot of the toxicology work who's done a lot of good stuff too. Ah and I think, yes, I think by and large, they've done a decent job in the field.

(Science Editor/Correspondent D, UK-Quality Newspaper)

Here, then, these journalists suggest a slightly different picture of source–media relations than that which had otherwise occurred in recent scientific controversies. This is important as the nature of the sources is likely to have a significant influence on what types of nano 'imagery' will circulate in the media. As Wood et al. (2007) highlight, whilst some radical versions of nanotechnology have largely lost sway in recent years, the technology's proponents (Drexler, Kurzweil etc.) have had an opportunity to build a media reputation during the corresponding time period. Similarly, Losch (2006) observes that though many of those working in the field have since distanced themselves from fictionalised, futuristic images of nanotechnology, they remain popular with the media. Those stakeholders and radicalised images of prominence in the initial framing of the topic have the ability to remain prominent regardless of shifts in how the topic is covered. Communicating these radicalised images of nanotechnologies at the outset, coupled with the 'blank slate' nature of many public attitudes towards the fields, has created an opportunity to strongly influence media coverage and thereby, at least potentially, public perceptions (Gorss and Lewenstein, 2005).

In the UK context this has allowed for certain scientists to enhance their media image while gaining experience as sources during a period of relative media optimism surrounding the technologies. This may influence, in turn, wider discourses about the risks and benefits of nanotechnologies across the media realm (Nisbet and Lewenstein, 2002). Many of the radical versions of how nanotechnologies will look and impact on us in the future have been traced back to the public discussions of specific personalities working in the field. Where these stories are then picked up within media coverage, a more obvious participatory relationship exists between the journalists reporting on nanotechnologies and scientists seeking to secure its future applications. Scientists have 'come out fighting', to 'sell' their subject, by providing newsworthy examples for the media, full of imagery and metaphorical opportunities, and some journalists have clearly picked these up. For these science and environmental correspondents, then, there has been a marked change in the more proactive stances of some scientists involved in the nanotechnologies field. Although pressure groups and companies were also mentioned as potential sources, this seemed tempered by a dose of scepticism:

> With nano the sources are mostly from the university institutes and the research councils. In terms of pressure groups I don't pay a lot of attention to Greenpeace on it... pressure groups are often using the same sources as government, often the Royal Society scientists. Having done quite well on the GM campaign, the pressure groups were looking around for a cause to take up. But it's not clear that they are that frightened of it. It's not clear where the alarm is coming from.
>
> (Science Editor/Correspondent A, UK-Quality Newspaper)

> Well I mean I'd rather, you know, ah try and translate from what a scientist tells me what's going on than from, you know, a company or a pressure group. I think in general they tend to be a bit more sort of informed about what the reality is. But I mean when you say what role do they play I mean it's not like scientists are beating a path to my door to talk about it, but I do get a fair amount of press releases.
>
> (Science Journalist B, UK-Quality Newspaper)

In these extracts journalists appear to remain most trusting of the scientists they speak with directly, despite the increasingly complex relationships between industry and science. In addition, their wariness of pressure group activities suggests that the caution accorded to nanotechnologies is potentially due to what may be perceived to be ulterior interests and/or motivations.

Finally, one interviewee returned to the involvement of Prince Charles. When asked about the shift in coverage from science, technology and environmental pages following the Prince's intervention, the journalist reflected on some of the difficulties this entailed:

> Well, that's actually an interesting point. You know this is where you sometimes get disasters like, dare I say it, *The Guardian* and the Pusztai affair handled by general correspondents who

didn't ... who sort of took a naïve, scepticism free line on what they were being told and the story was blown up very big and, you know, I think sometimes ... I mean particularly the grey goo scenario, you know, if you talk to anyone who's familiar with, you know, who's actually thought about making nanoscale replicators, it's just not on basically for various reasons. So it's a bit scary to think that in that case, you know, someone has swallowed the Prince's anxieties hook, line and sinker without thinking, hang on, what do the real scientists think about this stuff? Of course, the Prince, you know, he positioned the story in a clever way so he could claim that it was all those ridiculous journalists that hyped it up at the end. But that's the sophisticated game that, you know, he plays with the media.

(Science Journalist B, UK-Quality Newspaper)

The topic of nanotechnologies, including their definition and news appeal, offered us an interesting point of access by which to explore with our journalist interviewees certain key issues in the news reporting of science. The fact that nanotechnologies attracted coverage at all can be explained by the topic's obvious appeal for the journalists who wrote the stories. That is, its 'newsworthiness' derived, in part, from either extreme (Drexler) or fictional (Crichton) visions. However, the interviews suggest that a range of factors are likely to shape the actual nature of coverage, including views on the credibility of particular sources and the perceived social role of the media. As the interviews also made clear, journalists do not see themselves as science communicators, but are nevertheless acutely aware of their own responsibilities to 'the public' for whom they write and the news organisations for whom they work. That is, like scientists, they have a clear sense of their own role, and similarly subscribe to the view that coverage should be and can be 'accurate'; hence the concern to consult the 'experts'. However, as we found for the period of our study, along with stories of 'science facts' and commercial applications, news articles on nanotechnologies convey a strong element of fictional imagery and focus on celebrity figures, thus conveying a mixed portrayal of nanotechnologies. This raises the question – to which we turn in later chapters – of the responsibilities of scientists themselves in generating particular imagery through their explanations of innovations.

Conclusion

Our knowledge of the framing of a diverse range of nanotechnologies is developing as they are being introduced and integrated within societies. This study was limited, in that it did not examine audience's reactions to the media coverage of nanotechnologies. However, as we will explore in the next chapter, it is clear that research on public perceptions of the field is growing. This study's focus on UK-based coverage would further benefit from comparative research concerning the news framing of nanotechnologies in other national contexts (especially in the developing world), as well as by investigations into corresponding representations across a wider array of media genres (including digital and Internet based-formats).

In the UK context, this study demonstrates that during the formative period of news coverage on nanotechnologies, the subject was largely covered by the elite publications. As has been found in US research (Gorss and Lewenstein, 2005; Stephens, 2005), it was dominated by coverage supporting a scientific innovation or business focus with an optimistic tone driven by news sources largely from the scientific community. Although news stories on nanotechnologies were framed in a variety of ways during the period of our study, the predominance of science-related frames across different national newspapers was undoubtedly due to the disproportionate influence of scientific researchers as news sources. Perhaps this narrow range of sources, when coupled with explanatory frames that lent themselves to elite rather than popular forms of news coverage, further helps to explain why nanotechnologies did not attract more sustained forms of coverage across the period.

Journalists provided a significant dimension to this study, offering an insider's perspective on news coverage in the UK context. The choices they made in how they reported on this issue, the sources they used and their perspectives on the risks and benefits of the field were often couched in the language of 'common sense' notions of day-to-day reporting. Various UK events and source interventions had provided politicised, partisan or 'celebrity' news hooks for coverage, frequently drawing attention to areas which journalists swiftly argued lacked credibility, such as allegations about 'grey goo'. Further such coverage drew attention away from risks, such as those around nanoparticle safety, which were perceived as having greater feasibility, according to some journalists. For them, and despite their personal attention to the 'particulate issue', there are practicalities of reporting on it at this time. Nanoparticle safety is an area of long-term risk and current scientific uncertainty, and as such has remained relatively under-explored. Despite scientific and journalistic awareness that it may be an area which generates news coverage in the future (see Wilkinson et al., 2007), we cannot say from these findings alone if journalists are deliberately striving to avoid generating unnecessary alarm amongst the public or criticism from the scientific community. In any case, by not covering this issue in greater depth, they appear content to play a waiting game for more extensive research to emerge in the future.

At a time when policymakers are focusing their efforts on public perceptions of nanotechnologies, there seems to be a certain degree of complacency about the role and significance of the news media in forming public knowledge. The media are one of the few sources of public knowledge on nanotechnologies, and their specific framing of issues following particular 'events' cannot be taken for granted. One can only speculate about the media's impact on shaping the agenda for public debate should nanotechnologies establish a more prominent position in future mid-market or red-top press coverage, or be depicted in a more pessimistic tone as a greater range of sources become politically engaged in media discussion.

5
Nanotechnologies, Public Knowledge and the Media

Where do publics derive their information about nanotechnologies? How may the ways in which information is presented in the media and other forums potentially shape public knowledge of and responses to this field? The question of how technologies are represented and how this may affect public responses and ultimately public trust has become a key issue in the contemporary governance of new technologies. Publics' reactions to technologies – whether they will support or reject them – is seen to be shaped not only by the substantive content of information but also by how that information is 'packaged' and transmitted. The recent history of technology controversies has alerted scientists and policymakers to the power of 'public opinion' to influence the policies and regulations governing technologies. Adverse media publicity, it is recognised, can quickly derail what are seen to be promising new developments and undermine trust in authorities.

Reflecting a growing recognition of the importance of 'engaging the public', especially during the early phase of technology development, a number of new initiatives have been undertaken in the field of nanotechnologies in recent years. This chapter examines the rise of concern about 'public engagement' in the nanotechnology field in the UK and a number of other countries, and what this has meant in practice thus far. Next, it goes on to consider the role of the media in the formation of public knowledge about these technologies. As we argue, the question of what 'public engagement' means in practice is far from being settled and there are many uncertainties about the impacts of particular initiatives. Further, despite widespread acknowledgement of the potential significance of the media in influencing views and establishing the agenda for debate, the role of the media in forming public knowledge has been poorly theorised. Scientists often blame 'the media' for 'misrepresenting' nanotechnologies, as well as for not adequately conveying 'the science facts' in an effective manner. We suggest that this reveals a simplistic portrayal of science mediation and denies the power relations of science.

Our discussion begins by exploring the context for both surveys of knowledge about nanotechnologies and public engagement activities in the UK and internationally. It then considers some of the factors that may shape how news stories on nanotechnologies are framed. Drawing on data derived from a study of the production of news on nanotechnologies carried out by the authors, it examines how scientists view the relevant news coverage and what can be learnt from this about the current nature and future prospects of communication efforts. As we argue, the communication of knowledge on nanotechnologies is a complex and ambiguous process with uncertain impacts. It is virtually impossible, after all, to show a direct affect between communication efforts and the formation of public knowledge. Still, we shall identify and discuss several reasons why any serious effort to better understand the formation of public knowledge about nanotechnologies in the future will need to pay greater attention to the workings of the media.

A climate for public engagement

In recent years, the attitude of 'the public' to nanotechnologies has been an area of considerable research interest across the UK, Europe and the US. Following the example set by the Human Genome Project and other new genetic research, early investigations have been recommended for the exploration of the ethical, legal and social implications (ELSI) of developments. ELSI programmes have become a standard feature of many large-scale technology initiatives, often comprising a proportionately small but still sizeable aspect of their budgets. For example, with the Human Genome Project, the US Department of Energy and the National Institutes of Health devoted 3 to 5 per cent of their annual HGP budgets, respectively, to the ELSI issues surrounding genetic information, which is presented as a model for ELSI programmes around the world (Human Genome Project Information, 2008). Concerns have been voiced by many individual scientists and science policy groups about the potential for public responses to nanotechnologies to replicate those evident in relation to some genetic innovations, for example, GM crops and food and cloning (Turner, 2003). Public responses as well as physical dangers have come to be perceived as a risk associated with technology development that need effective management (see Corrigan and Petersen, 2008). The trajectory of nanotechnology innovations (as well as the responses to them) has varied across jurisdictions, however. In the UK and many countries in Europe there has been a tendency to take a more precautionary approach (Clift, 2006; Hunt, 2006). Meanwhile in countries such as Japan, there has been a low level of concern about the social implications of new technologies such as nanotechnologies, which are largely regarded as devoid of human values (Masami et al., 2006).

Beliefs about the benefits of technologies and how they may advance economic, health and social goals, and about the role of the state in regulating associated risks, are inevitably shaped by national histories, local politics and value systems. In many countries, arguments for the development of nanotechnologies and other new technologies often utilise the rhetoric of nation-building, making reference to public benefits and citizenship responsibilities (Jasanoff, 2005). This is evident, for example, in the UK government report, *Too Little, Too Late: Government Investment in Nanotechnology*, published in 2004. It noted the country's 'failure' to maintain its 'prominent position in the field', achieved in the 1980s, and expressed the desire to make it 'the major player in nanotechnology that our earlier progress in this area should have produced' (House of Commons Science and Technology Committee, 2004, 3–4). In countries with strong technology-based economies such as the UK there is likely to be much at stake in the development of nanotechnologies, especially where economic prosperity, health and social welfare are seen to be potentially jeopardised by what is seen to be an 'unaware', 'ill-informed' or fearful public. The challenge for policymakers in such cases is to strike a balance in supporting promising new technology developments and ensuring that associated public concerns, particularly about risks, are seen to be adequately addressed.

In the UK, the potential impacts of nanotechnologies and their social reception first gained prominence when the joint RS/RAE study, *Nanoscience and Nanotechnologies: Opportunities and Uncertainties*, was commissioned in June 2003 (RS/RAE, 2004; see also Chapter 3). This followed the precedent set in the US where the ELSI implications of nanoscience had been under discussion since the beginning of the decade, largely under the framework of the National Nanotechnology Initiative (Roco and Bainbridge, 2001; Sandler, 2007; Schummer, 2004). The RS/RAE enquiry was novel, bringing together two scientific academies and an extensive working group to explore the broad scientific and technological fields, alongside a consideration of the ethical and social issues. However, initially there was also apprehension that such an enquiry could be motivated to stifle rather than support widespread debate (Rogers-Hayden and Pidgeon, 2007). Similar concerns around the organisation and methods of such approaches have been raised by prior attempts to explore the social impacts of new genetics (Cranor, 1994) as well as attempts to encourage dialogue on subjects such as genetic modification (Gaskell, 2004). The fear was that scientists, academics and interest groups may set the research agendas under the guise of openness, whilst concurrently providing a PR exercise for the scientific endeavour undertaken by suggesting proper attention is being paid to its ramifications (Cranor, 1994).

The Royal Society/Royal Academy of Engineering study, which has provided something of a yardstick in many recent science and policy debates about nanotechnologies, reflects some of the concerns about the public definition of nanotechnologies. Importantly, it added credence to the idea that 'nanotechnology' is a contested concept. Recognising the wide range of disciplines and diverse applications encompassed by the field, the report suggested a pluralised use of the term. Surveying the present and future applications of technologies, the study highlighted promise in the production of cost-effective materials but vetoed the technological possibility of self-replication in the foreseeable future. In terms of health and environmental aspects, it became particularly focused on the safety of nanoparticles, with risks to those working with nanoparticles identified as being most pressing. However, there was recognition of uncertainty surrounding the risks posed by nanoparticles released into the environment or used in products. As well as recommending a responsible approach by manufacturers, the report strongly suggested that an interdisciplinary centre should be funded by the UK Research Councils, in order to broach the wide gaps in knowledge suggested. From a social and ethical perspective the report recognised significant difficulties in predicting which areas may create public concern. Issues of control and benefit were identified as key factors, as were nanotechnologies' contributions to human enhancement in the future, with the suggestion again that this warranted research via the Research Councils. Finally, the Report suggested all regulators, and specifically the Health and Safety Executive, put into place forward looking review procedures in the light of the developments, with a particular scrutiny placed on any areas which could fall outside of the existing regulatory boundaries (RS/RAE, 2004).

Following the publication of the joint RS/RAE report in 2004, the government has been engaged in attempts to implement many of its suggestions. The government identified its response to nanotechnologies as a key aspect of a broader attempt to 'make substantial and sustained progress towards building a society that is confident about the governance, regulation and use of science and technology' (HM Government, 2005a, 3). This commitment was reiterated not only in its supportive tone to many of the RS/RAE's key suggestions but also in its collection of the dialogue baton, so that the report would form the first stage of a continued 'dialogue to enable both the science community and the public to explore together both aspirations and concerns around the development of nanotechnologies' (HM Government, 2005a, 7). Furthermore it highlighted the diversity and development stage of the technologies and provided an opportunity for the responsible expansion of nanotechnologies. Aligned to this it suggested the need to fill knowledge gaps around the properties of novel particulates, and that a precautionary approach would be taken to regulation. However, in the two years following this initial response, the government faced mixed reviews regarding its progress (CST, 2007). In particular it has been criticised for the proportion of funding contributing to developing new applications, in comparison to exploring potential health, safety and environmental hazards, and likewise the failure to sufficiently close knowledge gaps, to ring fence funding or to provide a dedicated research centre (CST, 2007; RS/RAE, 2007).

The next GM?

In the UK, then, lessons derived from the reception of other technologies have clearly influenced thinking around the development of nanotechnologies. In particular, concerns have been raised about the potential for nanotechnologies to be 'the next GM' and to be subject to a public backlash similar to that experienced with biotechnology (Mayer, 2002). Comparisons have been made internationally to the handling of the two technologies (Kulinowski, 2004; Mehta, 2004; Priest, 2006; Sandler and Kay, 2006). Gaskell et al. (2005), for instance, draws many comparisons between developments in biotechnology in the 1970s and those that are being witnessed with nanotechnology today. In particular, parallels have been drawn in relation to the potential for negative public attitudes to develop and to the manner in which scientists are applying a self-imposed moratorium, considering ethical, legal and social implications as the field progresses (Gaskell et al., 2005).

Despite early attempts taken within the UK to study the implications of nanotechnologies, some policymakers are concerned that a GM-style public reaction may still occur. In a 2007 report commissioned by Department of Environment and Rural Affairs (DEFRA), which explored the potential environmentally beneficial nanotechnologies under development, the comparison to GM was again made (Walsh, 2007). The report warned that support for nanotechnologies, as indicated in current research into public views, could rapidly shift in the UK context where attitudes to GM, risk and trust had been affected and the positive opportunities for GM to reduce CO_2 emissions were ignored (Walsh, 2007). The report suggested that in the event public concerns around safety arose, they were likely to be around toxicological impacts. In this way, it echoed a typical policy-making assumption, namely that 'open, informed debate on the issues surrounding nanotechnology is the most effective way to ensure that there is public support for the development of this new technology' (Walsh, 2007, 80). Noteworthy here is a discourse of uncertainty. Despite the considerable attention being paid to communicating and learning about public attitudes to nanotechnologies, there remains a sense amongst experts that at least one aspect of the technology *will* become controversial. What aspect that is, or how it will be controversial, remains as tentative as the future of the technologies themselves.

Einsiedel and Goldenberg (2004) have warned that such comparisons to the biotechnology experience have been misinterpreted, bringing simplistic responses to discussions, and provoking a proactive PR exercise to 'educate' the public in order to avoid hyping claims. They suggest this has drawn attention away from the real issues – including addressing the broader risks alongside the benefits, acknowledging issues of ownership and control and recognising the importance of public trust not just the communication of information (Einsiedel and Goldenberg, 2004). By focusing on the specific, yet still uncertain, impacts of nanotechnologies, the broader picture of what went wrong in relation to the biotechnology debate may be neglected. Sandler and Kay (2006) similarly caution that the GM analogy may be unhelpful, as it overstates the potential for the technologies to be rejected when they have quite different characteristics.

Furthermore, what we have described is not limited to the UK. In Europe and the US policymakers have also learnt from past mistakes with biotechnologies. A cautionary approach has been mooted in the US, where assessing public attitudes to nanotechnologies has been more strongly modelled on the ELSI approach taken by the Human Genome Project, and the response to other technologies such as stem cell research has been examined for pointers on handling the reception to these new technologies (Schummer, 2004; Tourney, 2004). In some European countries there have already been instances of concern about products containing nanoparticles or nanomaterials. In Germany, for example, fears about the health implications of 'Magic-Nano' have been voiced, despite the fact that the product did not contain nano-sized particles. These have been mediated by a range of policies encouraging public dialogue (Bowman and Hodge, 2007). The significant difference in the UK has been the particular style in which public views have been incorporated into discussions.

From the deficit model to 'upstream' engagement

At the same time as products comprising nanomaterials are beginning to appear on the shelves – it has been estimated that there are already 500–700 commercially available nanotechnology-based consumer products on the market (Woodrow Wilson Project on Emerging Technology, 2007) – scientists and policymakers are advocating early or 'upstream' public engagement. Such engagement arguably allows the input of public views before crucial research and development decisions are made (Rogers-Hayden and Pidgeon, 2007). In official rhetoric, this represents a shift from the earlier so-called 'deficit model' of public understanding, which assumed that the task of science communication was to raise awareness and understanding of issues among an assumed unaware or 'ignorant' public. While the 'upstream' approach has gained widespread currency since the publication of the RS/RAE report, recognition of the need to move away from a public deficit model towards a model premised upon broader public input has been recognised since the UK government's influential *Science and Society* report, published in 2000 (HMSO, 2000; Rogers-Hayden and Pidgeon, 2007). The aims of this shift in policy emphasis have been articulated in a recent government report:

> The focus of the Government's Science and Society public engagement activities has moved forward from simply promoting public understanding of science to the wider agenda of facilitating public engagement with science and its application. This has the aims of: government and scientists responding proactively to public priorities and concerns; people having greater confidence in the benefits offered by science; greater engagement with major issues facing society, such as climate change; and careers in science becoming more attractive to both adults and children.
>
> (HMT/DfES/DTI, 2004, 103)

The movement towards 'public understanding of science', originating most strongly in the 1985 Bodmer report in the UK, was hindered by a series of problems, not least that of defining any of the three concepts it entailed: 'public', 'understanding' or 'science' (Rose, 2000; Royal Society, 1985; Turney, 2002; Wynne, 1995). Rather than opening science up to democratic involvement, the movement and policy initiatives sought to define appropriate questions, concerns and issues with a primary reliance on scientific expertise (Collins and Pinch, 1998; Hilgartner, 2000; Irwin, 1995). Within their 'deficit' frameworks, which linked factual knowledge to understanding, some approaches to science communication assumed that understanding and education about science would result in support for technologies (Evans and Durant, 1995; Gregory and Miller, 1998; Irwin and Michael, 2003; Michael, 2002). Recognising the diversity of public views, science and society programmes have attempted to move away from traditional styles of public understanding with the aim of being more deliberative and inclusive in their approaches. Interaction with 'the public' takes a range of forms in these approaches with methods like deliberative opinion polls, citizen juries, consensus conferences, Internet dialogues and focus groups (POST, 2001).

According to its proponents, 'upstream' engagement involves the consideration of a technology before significant research and development decisions are made with the aim of moving towards a richer public discussion about the 'visions, ends and purposes of science' (Wilsdon et al., 2005, 34). Upstream engagement, however, is often oversimplified; it is not meant to imply merely discussion occurring in advance but technological development being informed by public views (Rogers-Hayden and Pidgeon, 2007; Wilsdon et al., 2005). In the UK, a number of nanotechnology public engagement initiatives have followed in the wake of the RS/RAE study. This has included the creation of the Nanotechnologies Engagement Group to foster links and learning between public engagement projects focused on nanotechnologies and SmallTalk, a series of events focused on nanotechnologies across the UK (Gavelin et al., 2007; Wilsdon et al., 2005). At the European level activities have included NanoDialogue, a consortium of scientific, participation and communication experts who created a modular exhibition, a series of participatory events and a survey to capture and encourage public discussion of the topic (Heckl, 2007).

The review of governmental progress since the 2004 RS/RAE report suggested that while there had been extensive UK engagement with 'the public' involving programmes such as 'Nanodialogues', 'Smalltalk', 'WIST' and 'Nanojury UK' the channelling of these findings into policy initiatives has been unclear (CST, 2007). The impacts on policymakers of two-way approaches to communication is of increasing concern (Rogers-Hayden and Pidgeon, 2007); however it may often be difficult to trace tangible suggestions made in public settings to direct policy implications. Public engagement has come to dominate the social science focus (Wood et al., 2007). Schummer

(2004) suggests that the debate around the social and ethical implications of nanotechnology is currently the strongest, even perhaps
only, impact that nanotechnologies are having on society. Others
have suggested that this context has encouraged a distortion of the
debates that have occurred, the focus on avoiding a GM-style scenario subtly influencing the objectives of some public engagement
activities to avoid a backlash alone rather than fully utilising the
opportunity to debate substantive issues (Sandler and Kay, 2006).
A further criticism is that there are few realistic scenarios thus far
to present for discussion (Sandler, 2007):

> The dominance of the radical perspectives in the initial debate
> surrounding nanotechnology meant that it appeared in danger
> of becoming polarised before it had been allowed to mature...A
> consequence of this is that the focus of discussion, at least in the
> UK, became the issue of social acceptance and citizen's participa-
> tion in science and regulation. The question of how and in what
> way nanotechnology would develop, in the absence of this, was
> either taken for granted or ignored; an effortless view of science
> and the non-problematic view of technology transfer apparently
> being adopted, as the issue became one of influencing the goals of
> the scientist rather than how science and technology proceed and
> interconnect.
>
> (Wood et al., 2007, 6)

The focus on scenarios in public engagement efforts tends to direct
attention to utopian and dystopian visions of nanotechnologies,
which denies the complexities of the field and the considerable
uncertainties concerning the trajectory of technology development
(Wood et al., 2008, 19–20). It also suggests that meaningful engagement can occur when publics have very little starting knowledge of
nanotechnologies (Wood et al., 2008, 20). Given the currently low
level of public awareness of nanotechnologies it is difficult to see how
radical perspectives can be ignored as a basis for debate and action
when there remain few realistic scenarios available (Rogers-Hayden
and Pidgeon, 2007). Sandler (2007) also highlights that conversations
on social and ethical issues are appropriate at this stage as they can
highlight issues common to all technologies. He goes on to argue
that experts should not contradict themselves by suggesting some

areas of technological advance warrant investigation and public education whilst, at the same time, closing these same avenues to public discussion by stating they are unrealistic or exaggerated.

Disquiet over the impacts of upstream engagement reflects broader issues around the practicalities of engagement. Offering inclusive opportunities for publics to participate in decisions related to science and technology is now common across Europe and the US, but practical issues remain around when this should occur, what means it should take and how outputs are utilised (Evans and Plows, 2007). Many of the mechanisms such as referenda, focus groups, consensus conferences and so forth are under-researched, lack centralised monitoring and continue to reflect provider aims (Iredale and Longley, 2000; Irwin, 2001; OST and The Wellcome Trust, 2000; Rowe and Frewer, 2000). As Irwin and Michael (2003, 62) state 'although the case for public dialogue has been convincingly made, there is a substantial gulf between such discussion and particular examples of practical engagement'. In the US the approach taken by the National Nanotechnology Initiative, which has included consideration of both foreseen and unforeseen ethical, social and legal issues, has been criticised for focusing on environmental and health risks and public outreach and education, whilst promises of 'dialogue' with broader stakeholders remain elusive (Sandler, 2007). However, it does appear that nanotechnologies have provided a convenient access point by which to address this concern and offer comparative engagement strategies, although such new approaches will not necessarily avoid future controversy (Rogers-Hayden and Pidgeon, 2007), be they surrounding nanotechnologies or other areas of science.

A further question regarding upstream engagement pertains to its potential for manipulation, that is, to 'close down' rather than 'open up' discussion, should the opportunity for early engagement become the principal means by which public voices are heard (Rogers-Hayden and Pidgeon, 2007, 354). It may offer a legitimisation stamp to the onward development of technologies, however unpredictable their implications may be. Social scientists who are involved in 'engagement' activities run the risk of functioning as a kind of PR arm for business, either assessing the social impacts of nanotechnologies once the technologies are established or simply undertaking public engagement to satisfy demands for greater public involvement (Berube, 2006, 313, cited in Wood et al., 2008, 19). In the absence of substantial change in the power relations between experts and lay publics, there is the danger that public engagement strategies may reinforce rather than challenge the public deficit model (Wynne, 2006). As Wynne notes, the public deficit model of the public understanding of science has taken a number of different forms, and endeavours to 'know' more about 'the public' and 'its' views may perpetuate old assumptions under the guise of enlightened policy-making.

Gauging 'the public's' views on nanotechnologies

The new approach to engagement on nanotechnologies has been strongly shaped by concerns about adverse public reaction to the technologies (a related example is that of new biotechnology innovations, such as biobanks: see Corrigan and Petersen, 2008). Proponents of nanotechnologies recognise that should a public outcry occur it might have a significant impact on the success or failure of the numerous technologies and products under development (Cobb, 2005). The radical visions of 'nanotechnology' (i.e. portrayed as singular), which are based upon particular interpretations of the technology's applications, have tended to dominate public discourse (Wood et al., 2008, 21). This denies the uncertainties surrounding such innovations, as well as their reliance on other developments such as biotechnologies and information technologies (Wood et al., 2008, 21).

In conjunction with the more innovative engagement approaches designed to access public views, outlined above, more traditional methods for measuring attitudes such as surveys and interviews have been used. The deployment of these tools presumes a homogenous 'public view', one that may be measured and monitored over time. As surveys frequently provide the basis for policy decisions affecting the development and regulation of nanotechnologies, it is worth briefly examining some of their reported findings. Thus far, most surveys have been conducted in Europe, the US and Canada – reflecting what some commentators describe as a 'north-south' divide in the global development of nanotechnology innovations.

The reports on these surveys reveal that 'the public' currently has a low awareness of nanotechnologies at an international level despite their increasing prominence in a broad variety of products. For example, the Eurobarometer survey document, published in 2006, indicates that knowledge of nanotechnologies ranked lower than other technologies, such as biotechnology, wind energy and space exploration (European Commission, 2006). However, among those who had heard of nanotechnologies, it is noted that there is a higher level of optimism and a perception that nanotechnologies are 'low risk' (European Commission, 2006). The picture painted by the Euro-barometer data suggests that in the UK awareness of nanotechnology (44 per cent) is currently rated at about the same level of familiarity as gene therapy (43 per cent), although in comparison to familiarity with GM foods (92 per cent) the level of awareness remains pre-dictably low (European Commission, 2006). Despite the low level of knowledge expressed, in the European context optimism about nan-otechnologies appears to have increased since 2002, with only one in eight people being rated 'pessimistic' according to the Eurobarometer survey, with little notable gender or age variations (European Com-mission, 2006). This mirrors observed trends in attitudes towards biotechnology, which have improved considerably since the 1990s, with the level of optimism back to that which was last seen in 1991 (European Commission, 2006). In the UK context the most recent survey of public attitudes to science has similarly found a relatively low level of awareness or interest, with some perceived benefits and a low level of public 'worry' (RCUK/DIUS, 2008). All in all, the surveys present a generally optimistic view of 'the public's view', with Euro-peans perceiving nanotechnologies to be low risk, useful to society and morally acceptable (European Commission, 2006).

Like Europeans, US, Canadian and Australian citizens are also reported to have a low level of awareness and understanding of nan-otechnologies (Market Attitude Research Services, 2007; NSF, 2006; Priest, 2006; Scheufele and Lewenstein, 2005). The data, it is claimed, indicate that US citizens have generally been more supportive in their reception of nanotechnologies than their European counter-parts and that those who are aware of nanotechnologies are largely supportive (Cobb and Macoubrie, 2004; Gaskell et al., 2005; Priest, 2006; Scheufele and Lewenstein, 2005). Initial studies are seen to sup-port the view that, though there is recognition of the risks potentially involved, the benefits are seen as equal to or to outweigh them (Cobb, 2005; Cobb and Macoubrie, 2004; Currall et al., 2006; NSF, 2006). US attitudes are purported to be largely 'uninformed' but, as Scheufele and Lewenstein (2005) argue, this does not mean that exploring their attitudes is of little value, given that the history of public opinion research demonstrates that it is rarely a simple matter to assess an adequate level of knowledge on which an opinion is based.

The link between media coverage and public perceptions

Given the purported role of the news media in establishing the agenda of public debate on science and technology issues, it is surprising that few studies have directly explored the link between media coverage and public perceptions of nanotechnologies. In a number of science reports it has been suggested that the media may have a formative influence on public views on nanotechnologies and could be utilised in public engagement efforts (e.g. Currall et al., 2006; HM Government, 2005b; Nanoforum Report, 2004; RS/RAE, 2004). However, there has been a lack of empirical investigation of the current and potential future role of the media in the formation of publics' views. In broader examinations of public attitudes towards nanotechnologies, in particular where the current low awareness of the technologies has been explored, a notional media role has been identified. A recent report focusing on nanotechnologies by the insurers, Swiss Re, suggests that publics currently have a low perception of threat associated with nanotechnologies reflecting the low level of knowledge attached to them. Given this, there is the potential for media coverage to impact upon public perceptions and consequently, for this risk carrying insurer at least, 'nanotechnology will sooner or later emerge as a *public issue*' (Hett, 2004, 45, emphasis in original). Thus, the observation that public attitudes are positive but based on a low level of actual information offers little indication of any future public reaction. On the contrary, public responses may be more prone to the influence of negative or positive media coverage as their current views are based on such limited information (Cobb, 2005; Currall et al., 2006). Should a considerable amount of positive or negative coverage emerge, the media are likely to have a significant effect.

Jane Macoubrie found that the lack of information on the health and environmental risks of nanotechnologies and on the regulatory processes that will mediate them means US publics are often left to speculate on the impact of nanotechnologies, comparing them to past technologies such as asbestos, dioxin or Agent Orange

(Macoubrie, 2005). Scheufele and Lewenstein (2005, 660) suggest that, as is likely the case in many technologies but specifically nanotechnologies, publics operate under the 'Cognitive Miser Model'. That is to say, the majority of people, unless they have a specific expertise or motivation, do not use all available information sources to make decisions about new technologies. Instead, they take rational shortcuts, basing their opinions on their ideologies, religious beliefs and media coverage (Scheufele and Lewenstein, 2005).

Although there has been limited study of the specific impacts of media coverage of nanotechnologies on audiences, a number of US studies have focused on this question when surveying publics. These have found that the influence of the media on audiences is strong in the case of the emerging nanotechnologies if those audiences are the consumers of certain media genres. As seen above, publics currently have a low level of knowledge in this area, heightening the impact of any media influence in the future. In Scheufele and Lewenstein's (2005) telephone survey carried out with over 700 US citizens, it was found that use of the mass media was one of the strongest influences on responses to questions about nanotechnologies. More significantly, the survey suggested that heavy users of science media (including TV, newspapers or the Internet) were more likely to assess the benefits of nanotechnologies as outweighing the risks. Further, it was found that these assessments correlated with a positive media framing of the area in the US at the time of the survey (Scheufele and Lewenstein, 2005). Macoubrie's (2005) discussions with 12 groups of 'informed' private citizens ($n = 177$) in the US found that although participants had a generally low appreciation of nanotechnology, the media constituted a significant source of information. Twenty-six per cent of participants who cited a source of knowledge about nanotechnologies claimed that they had heard of the technologies via public or commercial television and radio, 17 per cent had been informed via magazines, 12 per cent by science fiction and 10 per cent by newspapers (Macoubrie, 2005). A total of 16 per cent of participants responded that they had heard about nanotechnologies via word of mouth (Macoubrie, 2005).

Whilst Cobb (2005) could not measure the impact of specific coverage on public views, he sought to examine the potential impact of nanotechnology framing via the use of a framing experiment presenting both positive and negative information about nanotechnology. By creating scenario- type information, each centring on a different framing condition such as health risks, he was able to elicit some interesting findings regarding the influences of portrayals on views. Cobb's (2005) results suggested that one-sided frames, those which highlighted a specific risk or benefit, were the most influential, whilst two-sided frames where participants were exposed to two sides of a debate were less influential. A further interesting finding emerged around the issue of trust in industry: when presented with a risk scenario, respondents' trust in industry declined, but when reversed, with beneficial scenarios described, participants demonstrated no increased trust in industry (Cobb, 2005). However, the study also identified few dramatic changes in perceptions amongst participants, suggesting 'we should not expect a sudden or dramatic shift in public opinion' on the basis of the extensive range of risk and benefit scenarios proposed (Cobb, 2005). Although the study has weaknesses, including the contrived nature of the frames and the selection of limited risks (Cobb, 2005), it is noteworthy that it was the one-sided frames which proved most influential. This invites the speculation that it may be these latter frames that emerge most strongly in the event that a significant public issue occurs.

In the European context where engagement models have increasingly been favoured, there are also indications that the media may be an important source of public information on nanotechnologies. In the European NanoDialogue project the reality of fictional imagery was one of the first issues to be raised by focus group participants, who also suggested that the media were important vehicles for the dissemination of information on the topic (Heckl, 2007). Though this suggests a slightly reductionist conception of the role of the media as a format for popularisation, it again points to the potential influence the media may have where nanotechnologies are concerned.

In broader terms the impact of debates concerning the future of nanotechnologies on science itself has been questioned. Wood et al. (2003) query whether the 'grand visions' associated with the more radical conceptions of nanotechnologies (and arguably those which have often attracted media attention) will shape their future development. Schummer (2004) similarly traces some existing scientific depictions, metaphors and concepts used in the nanoscience field to historical science fiction, which has traditionally used the nanoscale as a story component. The link between fiction and non-fiction broadcasting and public knowledge of and responses to nanotechnologies therefore warrants further examination. As with the portrayals of stem cell research and other emergent biotechnologies, the depictions of nanotechnologies involve the blurring of 'fact' and 'fiction' that may make it difficult for audiences to assess the significance of reported developments (Petersen et al., 2005).

Scientists' views on the media

The media coverage of science has long been a controversial issue for scientists. Criticised for reporting on science negatively and over-representing minority views, the media have long been distrusted by scientists (Ali et al., 2001; Bartlett et al., 2002; Kitzinger, 1999; Stocking, 1999). Conversely, coverage and journalistic angles that are too optimistic are criticised for hyping and sensationalising, often following the heralding of the latest medical cure (Nelkin, 1987). What is common amongst attitudes to both negative and positive depictions is the criticism that media coverage lacks sufficient detail, be it methodological content or qualifiers (Evans and Priest, 1995; Gunter et al., 1999; Nelkin, 1987; Pellechia, 1997). At the basis of many disagreements among scientists and journalists, however, is the absence of a shared view on the goals of the media. While, for journalists, the role of the media is to provide information, entertainment and criticism, for scientists, it is to provide public education and academic content (Reed and Walker, 2002). This absence of an agreed position on the role of the media has fuelled mutual suspicion and mistrust.

Despite such mistrust, in recent decades a growing number of scientists have acknowledged the importance of engaging with journalists to improve the image of science. Scientists have come to recognise the value of public visibility, of marketing themselves to raise the profile of their work in order to win government support and secure funding (Dunwoody, 1999; Nowotny et al., 2001; Paul, 2004). Many scientists have expressed concern about a lack of public confidence in their expertise. Surveys of scientists have suggested that they perceive members of the public to be more trusting of media portrayals than of themselves as experts (MORI, 2000). To improve interactions between scientists, journalists and ultimately media consumers, more science organisations (e.g. the UK's Royal Society) are investing in media and communications training, or using organisations like the Science Media Centre based at the Royal Institution in London to support scientists through interactions with the media (Royal Society, 2007; Science Media Centre, 2002). Likewise, reporters are undergoing scientific training, many of them working closely with scientists (Lewenstein, 1995). From the point of view of scientists, at least, interactions with the media have become an imperative; indeed a duty of citizenship.

A working group commissioned by the Royal Society (2006) produced a report suggesting that scientists have a responsibility to serve the public interest when the results of their research meet a number of criteria. This includes when research can further public understanding or participation in topical issues, publicise information which would impact on public well-being or safety, assist individuals in making informed choices where research may directly impact on their lives and, finally, aid the accountability and transparency of researchers, funders and employers (Royal Society, 2006). The report, whilst recognising complicating factors such as intellectual property rights, commercial considerations and security issues, states,

> In this context the research community has two main responsibilities. The first is to attempt an accurate assessment of the potential implications for the public. The second is to ensure the timely and appropriate communication to the public of results if such communication is in the public interest. These twin responsibilities should be embedded within the culture of the research community as a whole, and all practices should take them into account and respect them.
>
> (Royal Society, 2006, 5)

The report goes on to make a number of practical suggestions regarding how such research findings may be better communicated. It suggests that the cloak of anonymity surrounding peer-review might be removed, that the quality review of conference presentations be more transparent, that lay summaries and press releases should be better prepared, and that caution be exercised when publishing in open access formats online. (Royal Society, 2006)

Similar ideas have been articulated in relation to public engage-ment activities. Whilst those who do become involved in 'outreach' style activities often report that they are enjoyable and that com-municating with the public is not as difficult as they expected (Lewenstein, 1995), many institutions are still in a period of tran-sition in regard to undertaking such activities. Scientists report that public engagement can be time consuming and that departmental support remains low. It is only recently that such activities have been 'sanctioned' more widely (Gunter et al., 1999; Hargreaves et al., 2003; Pearson et al., 1997; Peters, 1999). It should not be surprising, then, that scientists' expertise in public engagement, including media skills, is in the main under-developed. There appears to have been lit-tle concern to understand the day-to-day workings of the media and how this may shape the content and form of science news report-ing. Relatively little is known about how scientists actually interact with the news media, in general, nor about the ways in which scien-tists interact with journalists, in particular (Dunwoody, 1999; Gunter et al., 1999; Peters, 1995; Reed, 2001). Consequently, there has been a tendency to treat the media as a 'black box': to acknowledge 'its' distorting influence but to remain largely ignorant about 'its' inner workings (Petersen et al., 2008).

The tendency of scientists to 'black box' the media is confirmed by a study undertaken by the authors. As part of our effort to learn more about the mediation of knowledge about nanotechnologies, we investigated how UK-based scientists who worked in the nanotech-nology field viewed public engagement, that occurring both that occurring directly with citizens and via the media. An email sur-vey ($n = 37$) was carried out with scientists who had been quoted or cited in news items about nanotechnologies encompassed in the prior news analysis survey and with those who had been identi-fied as stakeholders in nanotechnology research and development. The latter group included those who participated in the aforemen-tioned RS/RAE study, academics who taught university courses in nanoscience or nanotechnology as well as scientists who had been recommended through these sources. In addition, a series of in-depth interviews ($n = 11$) were undertaken with a self-selected sub-sample of the surveyed researchers and academics. A more detailed analysis of the survey findings is reported elsewhere (Petersen et al., 2008; also see Chapter 4).

The survey and interviews explored a number of issues including the setting of news agendas, perceptions of nanotechnology coverage, the process of communication and the influence of stakeholder groups, alongside reflection on some of the more practical matters involved in interactions with members of the public or journalists. The survey and interviews were all carried out in 2005. The survey sample consisted of 21 Professors, 14 Doctors, one research student and a Chief Executive Officer of a research institute. Approximately a quarter ($n = 9$) of respondents had been working in areas related to nanoscience or nanotechnology for under five years, almost a third ($n = 11$) had worked in the field for six to ten years, while just under a quarter ($n = 8$) had been active for 11 to 15 years, and a further quarter ($n = 9$) for over 15 years. Most of those interviewed were senior academics (seven Professors and one Reader), a representative of a scientific organisation, a research student, and a Chief Executive Officer in a research institute were also involved respectively. In the next sections we will provide an overview of the key findings of this aspect of the study. In brief, the findings revealed some mixed views about the way the media portrayed nanotechnologies, especially the use of fictional imagery. The media's role in public engagement was acknowledged, with many expressing a desire to better utilise media forms to help explain nanotechnology innovations. However, given scientists' role as expert commentators and primary sources for stories on nanotechnologies, there was surprisingly little reflection on the complexities of news production, including their own role in framing issues. Indeed, in frequently drawing a distinction between news image and 'science fact', their comments suggested that they saw themselves as somehow standing outside the news production process.

Scientists' views on news coverage

The survey respondent's exposure to news coverage of nanotechnologies was found to be high. Only one respondent stated that they had not read any newspaper coverage of nanotechnologies in the two years prior to the survey occurring. A number of criticisms were made of coverage, including the use of certain language, metaphors and images, inattention to the important issues, and the confusing of 'fact' and 'fiction'. There was a general concern about the *accuracy* of science reporting. When elaborating on their views, scientists referred to the use of certain metaphors and anecdotes and suggested that sensationalised headlines created an inaccurate or 'misleading' portrayal of the field. However, responses were mixed, as is illustrated by the following quotes:

> Most stories result due to a breakthrough in science, which is reported but this is extrapolated to suit the sensationalism and sell papers. This popularises the subject but in the wrong way.
>
> (Professor, Technology Institute)

> The newspaper coverage with which I've been involved (*The Guardian*, *The Times*) over the last couple of years has generally been well-balanced and I've largely been impressed with the journalists' writing.... However, 'nanobots' always feature highly in press coverage, as do artists' renditions of Fantastic Voyage-like 'nanosubs' hunting down viruses in the bloodstream. I can understand why these concepts and images are continually used – as they certainly add a lot of interest to what might initially be seen as a dry science story. It seems that such 'sensationalised' images are used to 'spice up' otherwise well-balanced articles. Is this detrimental to science? It's a moot point – if the 'nanobot' or 'nanosub' image succeeds in attracting a reader's attention to a well-balanced and scientifically correct article then one might argue that the 'artist's impression' has served an appropriate purpose and this has been beneficial to science. If, however, the nanobot or nanosub image is the only information that remains with the reader, then this is extremely misleading and is rather detrimental to the future of nanoscience.
>
> (Professor, School of Physics and Astronomy)

The tension between recognition of the need to gain publicity for research, to provide a scenario for readers to envisage and the potential for the issues to become hyped was clear. In the view of one scientist, the press' tendency to 'over-sell' nanotechnology is no different from other areas of science, though at present the scientist recognised the general findings highlighted previously that nanotechnologies was receiving a generally positive response:

I find that most articles portray nanotechnology as a great opportunity for the future. Of course it gets oversold, but that's what journalists do – I do not think nanotechnology fares worse than other areas of science, in fact better, considering GM crops and gene technology.

(Professor, School of Physics and Astronomy)

According to a number of our respondents, 'sensationalism' or 'hype' may, in some instances, actually be *beneficial* for science. Many mentioned that for some science departments struggling to attract student numbers, hyped headlines and futuristic images could help draw interest in specific fields, although this was often followed by notes of caution:

Reporting just gets people's interest going. Many colleagues are worried about preventing bad influences on public opinion but in my view, it is impossible to strongly influence. [The] media just want good stories and sensationalism, good or bad, has to be there for current public taste. I think the idea of danger is actually GOOD to draw people into studying these subjects.

(Professor, School of Physics, emphasis in original)

I definitely share the view that nanotechnology has been hyped out of all proportion, which then means that the actual science cannot live up to the expectations of nanobots etc. However, almost any discussion of science in the media is a good thing.

(Professor of Organic Chemistry)

Grey goo and nanobots have of course a wonderful appeal in selling the more fanciful, and thus attention-grabbing, elements of nanoscale science. Indeed one might argue that as a strategy to encourage school children to become enthused by nanoscale science the prevalence of nanobots in the media might be a good thing. However, the distinction between that which is scientifically achievable now, and that which may be possible in the (near) future, and that which is simply misguided and scientifically incorrect must always be highlighted.

(Professor, School of Physics and Astronomy)

Although acknowledging that fictional imagery 'might be a good thing', these scientists did not acknowledge the role of fictional

imagery and popular metaphorical content in scientists' own representations of science (Brown, 2003; Petersen, 1999; Petersen et al.,
2005; Tourney, 2004). As Pense and Cutcliffe (2007, 353) state,
'nanoprose' is dominated by a sense of revolutionary proportions,
has an unlimited potential for change and suggests that the technologies will have an encompassing and permanent impact. References
to 'sensationalism' and 'misguided and scientifically incorrect', in
the above quotes, imply that a clear line can be drawn between 'science fact' and 'science fiction' and suggest that scientists have little
responsibility for metaphorical content and fictional imagery despite
sometimes seeing it as an opportunity to 'sell' the science. However,
there was a clear, if somewhat cynical, recognition that journalists
used such terms and narratives to attract attention.

> Journalists believe that the non-science educated public needs
> punchy (mainly doom-threatening) phrases to catch their
> attention.
> (Senior Fellow in Quantum Metrology and Nanoscience)
> Because we all like a good disaster story, especially those where
> hubris results in catastrophe (see Frankenstein.) Stories about ben-
> eficial uses of any technology, unless they involve curing cute kids
> or puppies, are fairly boring.
> (Professor, School of Chemical Engineering
> and Analytical Science)
> Grey goo has a great sound which people will remember and so is
> a nice handle to hang things on. Nanoscale robots fit nicely with
> science fiction medicine, so again something people can relate to.
> (Professor, School of Biological Sciences)

Despite a willingness to engage with the media and the varying
experiences attached to that (see below), the general impression of
media coverage was less than positive. Of the 36 respondents that
had read coverage, half ($n = 18$) described it as 'inaccurate' compared
to just over a third ($n = 14$) who claimed it was 'accurate'. While
almost three-quarters ($n = 27$) of respondents that had read coverage described it as 'sensationalised', only a quarter ($n = 8$) described
coverage as 'balanced'. Scientists were much more likely to describe
newspaper coverage of nanotechnologies as 'detrimental to science'
($n = 19$) rather than 'beneficial to science' ($n = 13$). Those who had

worked in the field for a lesser time interestingly had a more positive view of coverage, suggesting those who have longer experience in the field are more likely to hold a negative view of news coverage as they may develop investments in positive portrayals of nanotechnologies, sometimes challenged by news stories (Petersen et al., 2008).

Many of those involved in the survey had a realistic expectation of existing public regard for nanotechnologies, suspecting that generally the public lacks awareness of nanotechnologies, a view which supports the findings of a number of the studies of public attitudes we have previously outlined. Responses to both the closed-answer questions and the open-answer questions revealed a belief that the public lacked information about nanotechnologies. Approximately two-thirds ($n = 25$) of the respondents thought the public were 'uninformed' about nanotechnologies, while a smaller number ($n = 10$) stated that the public were 'very uninformed'.

Scientists' contributions to news coverage

Respondents' experiences as sources for media coverage were also relatively extensive and mostly described as 'positive'. Almost half ($n = 17$) of all respondents had been approached to contribute to or comment in newspaper articles, and many ($n = 14$) had gone on to do so. The survey did not examine reasons why three of those who had been approached had not gone on to contribute, but for those who had it was reported to be a satisfactory experience for the majority (11 out of the 14). Two participants had not been satisfied with the nature of the coverage in the stories in which they were involved, while another stated that they had been both 'satisfied' and 'unsatisfied'. Interestingly, of those approached, there appeared to be little connection between their desirability to act as a media source and the stage at which they were in their career: there was a relatively equal split in the length of time they had been involved in the area of nanotechnology. Five had worked in the field for fewer than five years, five for 6 to 10 years, five for 11 to 15 years and two for over 15 years. Thus, for this limited sample, experience and the length of association with the field did not bear a relationship to the likelihood of being approached as a source.

Some scientists were evidently more confident in working with the media than others and were encouraged by their universities to actively use the media to publicise their work. Others suggested they utilised the media to draw attention to recent research and funding:

> The exposure we had recently was, well we kind of touted it deliberately in a way because we wanted to, um, you know because of the internal politics of the university; we want a big splash for things that we do, especially if it's something that reaches the Vice Chancellor.
>
> (Professor, School of Physics and Astronomy)

Still others, however, described a broader 'duty' to engage the public, reiterating the policy lines on engagement which have clearly impacted on the work of scientists in recent years. This also parallels Rogers-Hayden and Pidgeon's (2007) findings that suggest that nanoscientists are generally in favour of a more involved public. For the following scientist, who worked for a scientific organisation, a sense of public duty and continuing funding were clearly concerns:

> I think on a basic level most science... is funded by the public, therefore it's our... a scientist's duty to make sure the public are aware of what is happening with their money and what the potential benefits are, I mean really in a self-interested way because perhaps if people aren't aware of what's happening they might not be so keen on having money spent in that way.
>
> (Representative of a science organisation)

When asked what would influence their decision *not* to take part in newspaper coverage, almost two-thirds ($n = 23$) expressed concerns about the quality of the information that they had been asked to respond to. A slightly lesser proportion ($n = 22$) held concerns about being misrepresented or misunderstood. Given these concerns about accuracy and being misrepresented, in some cases scientists sought to gain some control in their interactions with journalists by becoming 'proactive' as sources:

> We have a proactive PR policy and are relatively close (and always open) to the journalists we work with. We also run public

awareness events and also present to TV etc. and have literature explaining what our institution is about.

(Director, Research Centre for Nanotechnology)

Other strategies to attract media interest included developing websites, producing lay summaries and writing press releases, although it was stated that it could be difficult to predict when these would be picked up. Using organisations like the Science Media Centre was identified as useful, as was the availability of science writers in residence.

Increased funding for participation events and programmes was widely called for, as was the need for time and institutional support to carry out such activities, with the majority of scientists involved in the survey expressing positive views on public engagement. Only three survey respondents thought it was 'unrealistic' to engage with the public. Many responses suggested a willingness to participate in engagement strategies, and this belief was found to be especially evident among the more senior scientists; that is, those at professorial level. The scientists involved in the study clearly identified a role for the media in engagement in relation to nanoscience and nanotechnologies. However, as some pointed out, 'the media' is not homogenous and different kinds of media may call for a different approach:

I did interviews with a variety of people. I mean I did interviews on BBC radio, which was fine. I was on *Material World* on Radio Four, who were very good and surprisingly scientifically literate, I mean I was genuinely impressed, in the sense of that everyone I met, all of the producers, had PhD's, which I was quite taken aback by. So they were fantastic, and also I was on the World Service as well, and they were really good, they were fine. Then I was covered in America, mostly by scientific people with sort of scientific backgrounds but on the other hand, despite that they were very good as well, they asked intelligent questions, it was all entirely reasonable. In this country, so for example I was covered by the local press here, who insisted upon including comments on nanobots, tiny machines and that sort of stuff, despite the fact we explicitly asked them to take it out and they still put it in, and the level of scientific knowledge there was astonishingly poor.

(Professor, School of Chemistry)

For a number of scientists who are involved in media coverage, some forms of media provide a greater scope for control. However, as one scientist observed, the desire to change a story before it goes to press is likely to depend on the target audience:

In print you have a lot more control because generally in print it's a story that has been placed somewhere and you are working with science writers. They have an interest in getting the story right; you often see it before it goes out. You never see TV before it goes out. So, you see it before it goes out and if it's desperate to change it you can, but I never really normally do it. If it's wrong only the scientists will know about it and that is not the target audience. So I don't really care. I have a fairly relaxed attitude to it all.

(Professor, School of Physics)

Both rehearsing interviews and providing written materials for journalists were mentioned as methods to reduce the uncertainty of any encounters. In the case of nanotechnologies the novelty of the concept rather than the scientific fields it incorporates may lead scientists to feel nervous at the prospect of press attention. Scientists are likely to be concerned that current media discussion of the risks and benefits of nanotechnologies in a context of low levels of public awareness will help shape the conditions for public attitudes towards the field (Petersen et al., 2008). The scientists who were surveyed empathised with audiences who were trying to make sense of the challenging and emerging fields related to the technologies. One scientist, for example, stated that it was unfair to expect members of the public to understand the background of the area or to assess the experience and qualifications of those commenting in the media.

Scientists' perceptions of the risks and benefits of media coverage are likely to be influenced by their experiences of previous controversies, such as those surrounding BSE or GM crops (Petersen et al., 2008). However, as media attention shifts from one area of science, health or the environment to the next, scientists' levels of experience of and thus views on the media's reporting of particular issues are likely to vary. Our sample, which included many individuals who had been cited or quoted in news stories on nanotechnologies, may be atypical of scientists in being especially 'media savvy'. Whilst the majority of those questioned were acting *reactively* rather than proactively (Petersen et al., 2008), a number appeared acutely aware of the responsibilities and obligations of an ongoing active engagement with the media.

> If you want people to start seeing emerging patterns of why is this thing related to that other, you have to find a way to have more than punctual contact [with the media].
>
> (Research student, Materials Department)

> I mean there is only a certain amount of time I can put into it, and the real problem is if you stick your head above a parapet here they are desperate for people to talk to at the moment – they really want them – so it is either nought percent or one hundred percent and I can't deal with that.
>
> (Professor, School of Physics)

Contact with the media can be unsettling, especially as scientists may only have one or two media-generating incidences in their careers. The question of whether that situation will change as public engagement is encouraged or whether it will be the more confident or high-profile areas of research that attract attention remains to be seen. Professor Richard Jones, who has become involved in a variety of communication activities and consultation processes around nanotechnologies, has described engagement with nanotechnologies as an unfamiliar and intimidating, though ultimately positive, process for scientists (Gavelin et al., 2007). He describes nanotechnology as a goal-orientated activity, and suggests that the goals are open to value judgements, rather than being simply a matter of technical decision-making. Further, as the science remains uncertain, discussing the future scenarios of this emerging science can make the science vulnerable to distortion (Gavelin et al., 2007).

Conclusion

Recent heightened science and policy interest in nanotechnologies has served to focus attention on how information about the technologies, including their risks, is communicated to publics. As we have noted, however, approaches to nanotechnology development have varied between countries, with some taking a more 'precautionary' approach to technology development. Being acutely aware of the significance of public responses to the trajectory of technology development, scientists and policymakers in these countries have sought to develop means of communicating with publics in order to ensure that public trust is not undermined and that potentially beneficial applications of technologies are not de-railed.

Public responses have themselves become a risk to be managed and consequently early or 'upstream' 'public engagement' has moved centre stage as an issue of technology governance. However, as we pointed out, current forms of 'engagement' are rather limited forms of communication and can be criticised on a number of grounds. Such efforts *in practice* do not challenge the power relations of science. Public engagement activities continue to be largely about ascertaining levels of people's awareness of issues and imparting knowledge, the assumption being that a more informed, technological literate 'public' will be more predisposed to supporting technologies. There has been little effort to understand how the news and other media construct risk and how experts' assumptions about nanotechnologies and 'the public' with whom they 'engage' may shape the communication process. Experts' assumptions about nanotechnologies, including their benefits and risks, and about 'the public' remain largely unexamined. As we have argued, the media have been a largely forgotten element in communication efforts.

Given scientists' frequent criticisms of media portrayals of science in general and nanotechnologies in particular, and recognition of the potential role of the media in establishing the agenda for debate, it is surprising that so little effort has been made by scientists to explore the workings of the media. Few appear to have paused to reflect on how their own role as expert sources and commentators may shape the communication of information. Their history of 'engagements' with the media had been largely about how to better exploit the institutions of the media in order to improve the image of science, rather than to better understand how their own assumptions and ways of working may shape their communications with journalists and the wider public. Our own research confirms that scientists have tended to view the media as a 'black box', in that while they acknowledge their importance in shaping views they fail to interrogate their routine workings. As outlined earlier, considerable expectations are attached to nanotechnology developments, with new applications expected in a range of fields in the coming years. The question of how technologies develop, and whether particular imagined applications are realised, will depend very much on whether publics support those developments and maintain their trust in the regulatory systems established to govern the impacts of developments. Given the histories of responses to technologies thus far, the fields of health and medicine, along with environmental sustainability, are likely to be key testing grounds for public responses to new nanotechnologies. With this in mind, in the next chapter we explore some of the promises and potential perils of these particular fields of application, showing how scientists themselves have sought to strike a balance in the representations of the attendant benefits and risks of nanotechnologies and pointing out how these may shape public views and responses.

6
Scientists' and Policymakers' Representations of Nanotechnologies

It is increasingly the case, as our discussion in previous chapters has shown, that the public perception of nanotechnologies is becoming a subject of considerable concern and action within science and science policy communities. More and more, scientists and policymakers have come to recognise that adverse publicity about and hostile reaction to technologies may work against the realisation of what they believe to be potentially valuable innovations. In the light of this, the discourse of early or 'upstream' public engagement, as articulated in the UK and some other countries, has emphasised the need for wide-ranging discussion about the nature and potential implications of technologies *before* they become established. This implies evaluation of any likely adverse affects of applications on human health, and on physical and social environments. A key challenge for proponents in this context, therefore, is communicating information to publics about nanotechnologies and their implications without generating widespread fears about unforeseen consequences.

This chapter focuses on scientists' and policymakers' representations of nanotechnologies and how they seek to communicate information about this field during a period of their relatively low – but growing – public visibility. In particular, it examines the difficulties they face in seeking to establish a positive portrayal of technologies when there is definitional debate about the nature of the field and uncertainty about the applications and implications of developments.

Nanotechnologies, as an emergent set of so-called 'enabling' technologies with an anticipated wide range of applications, would seem to represent a unique array of challenges for their proponents. Unlike fields of technology that have achieved a high level of public visibility, such as human genetics, there is no agreed language and repertoire of metaphors for representing this emergent field in the wider culture. Further, there are as yet few established non-expert communities of interest, like patient support groups, based on particular applications of nanotechnologies to generate and support shared values, narratives and strategies in relation to these technologies. The absence of an agreed vocabulary and non-expert communities of interest presents a potential challenge for scientists and policymakers who seek to communicate with lay publics. In particular, how do they establish and convey a positive portrayal of nanotechnologies in the light of the technologies' acknowledged uncertainties and risks?

In addressing this question, this chapter is based upon a study that examined the representations of scientists and science policymakers who themselves work in the burgeoning field of nanotechnologies.[1] Focusing on the fields of health and medicine and environmental sustainability, respectively, we highlight the ways in which our study's respondents gave expression to varied definitions of 'nanotechnology', as well as its perceived 'benefits' and 'risks'. While this study did not explicitly examine the media per se, its findings nevertheless complement our discussion thus far. Its participants, mostly delegates who had attended a scientific conference and members of a professional network (details below), agreed to help investigate a series of inter-linked questions through the use of online questionnaire surveys and in-depth interviews with a sub-sample of survey respondents who indicated a willingness (on the questionnaire) to contribute in this way.

Briefly, the study asked, first, how do scientists and science policymakers portray the future applications of nanotechnologies as applied to medicine and environmental sustainability? Specifically, in accounting for their views, how do they strike a balance in representing the benefits and risks for each area of application in order to maintain a positive conception of science? And, finally, what constructions of 'the public' and what 'it' needs to know are revealed by these accounts? While the study is based on a limited sample, comprising in the main scientists with backgrounds in chemistry, toxicology and risk analysis and research funders with an evident interest in the environmental impacts of nanotechnologies, it nevertheless aimed to generate useful insights into a number of communication issues relating to nanotechnologies.

Here we draw on the study's principal findings to highlight the nature of the challenge confronting scientists and policymakers who seek to communicate a positive portrayal of nanotechnologies to lay publics. As we argue, the particular way in which nanotechnologies have been framed, in terms of their 'benefits and risks', proves to be a delicate balancing act for nanotechnologies' proponents. We explain the nature of the challenge confronting scientists and policymakers in their communication efforts thus far, before proceeding to draw out some implications for future initiatives. First, though, in order to lay the groundwork for the discussion, it is useful to present some background on the socio-cultural and politico-economic factors shaping communication processes in this field.

The forging of a consensus around nanotechnologies

Nanotechnologies, as we have seen in previous chapters, represent an interesting instance of the forging of a broad consensus among scientists and policymakers about the significance of a field largely on the basis of strong *expectations* about their realisation. Technologies, no matter what the field, always embody expectations about their prospective uses, thereby helping to give shape to future applications. As Brown and Webster argue in relation to new biomedical technologies, the circulation of excessive expectations provides an important motivational force in technological development, serving to accelerate change and sustain investment (2004, 179). Often, the more uncertainties that surround a field, the more extreme are the promises. As Brown and Webster note,

> where technologies are highly novel there is the evident need to galvanize newly forming relationships, to encourage new interest and raise share value, etc. The more acute these uncertainties, the greater will be the need to draw on the motifs of revolutions, breakthroughs and radical change. So radical discourses about the future are often indicative of the early emergence of an NMT [new medical technology]. And, of course, the greater will be the likelihood that things will turn out far differently in the end.
>
> (Brown and Webster, 2004, 182)

Given the need to forge new alliances and investments in new technology innovations, it is hardly surprising that such technologies
are frequently surrounded by 'hype' (Brown, 2003). For example,
promoters of civilian nuclear power claimed during the early years
of development that the technology would produce electricity that
would be 'too cheap to meter' (Mehta, 2005), while promoters of
personal computers claimed that computers would save trees by the
electronic storage and coding of data (Hunt and Mehta , 2006, 279).
Similarly, nanotechnologies are surrounded with considerable hype,
often in the absence of substantial evidence of their promises being
fulfilled. Making much of the innovative aspects of technologies and
their revolutionary potential is part of this 'hyping' process. The question of whether nanotechnologies are genuinely innovatory is open
to debate and, as we show later, is even challenged by some scientists.
However, there seems little doubt that the radical visions are shaping
responses, both of the proponents and of the opponents, which may
affect how these technologies are eventually realised.

We have referred before to reports generated by governments and
science groups in this field which highlight the potential bene-
fits derived from nanotechnologies. In the UK, these include the
RS/RAE's (2004) *Nanoscience and Nanotechnologies: Opportunities and
Uncertainties* and the House of Commons Science and Technology
Committee's (2004) *Too Little, Too Late?: Government Investment in
Nanotechnology*. In addition, there are a growing number of nanotechnology conferences and industry events which trumpet new
developments in progress or 'on the horizon' (NanoForum, 2008).
Such forums serve as a means not only of sharing information and
promoting innovations widely among science communities but also
of raising awareness among policymakers, industry groups, NGOs
and broader publics about nanotechnologies' potential. Greatly
assisted by the Internet, proponents of nanotechnology innovations
have established links across disciplines and fields of practice at
the global level to lend the impression – if not the reality – of
a community of the likeminded who are pushing the frontiers of
knowledge.

Like population-wide genetic databases ('biobanks'), which have
become a global phenomenon, the field of nanotechnologies is being
cast as 'big science' involving large research teams and diverse actors
and networks spread across the world (Gottweis and Petersen, 2008).

It comprises various disciplines and experts engaged in diverse activities, including basic research, knowledge transfer and environmental management and assessment. Nanotechnology science and development has given rise to a huge industrial complex, spanning areas such as biomedicine, environmental management, the military, engineering, materials, mining and information and communication technologies. This industry is supported by a substantial infrastructure of public- and private-sector research programmes and funding initiatives; for example, the US's National Nanotechnology Initiative. Proponents of nanotechnologies, in particular scientists who are undertaking research in particular specialist fields and supportive science policymakers, constitute a marketplace of competing visions and promises of nanotechnologies.

Virtually all such proponents share the goal of seeking to 'sell' a vision of the benefits that will result from investment in particular innovations. These visions are promulgated to other scientists, policymakers and wider publics via various media, including press releases, where scientists frequently serve as expert sources, industry events and public education initiatives. The expectations of nanotechnologies are thus, in essence, a *social performance* – one that is sustained through a diverse array of activities for promoting narratives of where the technologies are heading and who will benefit.

The production of expectations

Nanotechnologies are products of particular times and places, expressing the interests and needs of the present, but with a view to the future (see also Brown and Webster, 2004, 180–181). These priorities continually evolve in line with broader changes in politico-economic conditions and relations of power. The history of biotechnology developments, perhaps most obviously GM food and crops, illustrates how the influence of different stakeholders may change over time and how this may affect the production of expectations and the realisation of technologies. Experiences of developments in the past often provide a reference point for imagining future scenarios, optimistic or otherwise. The expectations around nanotechnologies are multifarious and have been highlighted at various points in the preceding chapters. Most obviously, financial interests are crucial, since the economic 'spinoffs' from innovations are widely perceived to be substantial.

In 2007, a media release of the Australian Nanotechnology Alliance noted,

> A leading US research group estimated that goods including nano could be valued at $2.6 trillion, employing 7 million persons worldwide. Aggregating these figures into Australian data would see the sector at around $60 billion and employing 125,000 Australians. These figures indicate the impact of nanotechnology will be as large as information and communication technologies combined and more than ten times larger than biotechnology revenues.
>
> (Australian Nanotechnology Alliance, 2007)

In this press release, readers unfamiliar with the field of nanotechnologies are being encouraged to appreciate the extent of the potential economic benefits to be gained, not least by the comparisons made to other, more established fields of innovation. In this passage we see the language of expectations linked with the project of nation-building – the promise that the technologies will contribute to the wealth of the nation – seen also within the biotechnologies field (Jasanoff, 2005).

Typically, science and policy documents concerned with 'breakthrough' developments rarely acknowledge the potential exclusionary, discriminatory and oppressive implications of nanotechnologies; for example, the emergence of a 'nano-divide' within populations (Hunt and Mehta, 2006, 279–281). A focus on the technologies themselves and their potential has frequently served to narrow debate to a select number of substantive issues. Technologies are always embedded within social contexts and, as such, possess an inherently political dimension. They are always both enabling and constraining, invariably altering expectations in unanticipated directions, and in their applications always create the relatively advantaged and the disadvantaged, or 'winners' and 'losers' (Lehoux, 2006, 157–191). For example, the digital revolution has led to a broad division between the 'information rich' and the 'information poor' which is manifest in disparities in wealth and standards of living at national and international levels (the so-called 'digital divide'). Discussion of the potential for such uneven and discriminatory impacts fits uneasily with the officially preferred discourses of nanotechnology innovations, which tend to emphasise their expected beneficial applications based on dramatic, even utopian visions.

Many industries are keen to exploit the opportunities provided by the next technological 'breakthrough' and it is here that nanoscience is proclaimed to play a pivotal role. As noted in earlier chapters, nanotechnologies are expected to find application in agriculture, food processing, biomedicine, neuroscience, mining, the manufacture of materials and information and communication technologies. These potential applications are increasingly 'showcased' at events sponsored by industry or funding bodies who are keen to exploit the commercial opportunities 'enabled' by nanotechnologies, in particular as a result of their convergence with other technologies, such as biotechnologies, the neurosciences and information and communication technologies. For example, Nano2Life the 'first European Network of Excellence' funded by the sixth Framework Programme of the European Union (and involving 23 public or non-profit organisations from 12 European countries) hosts events on topics such as the applications of nanotechnologies in the neurosciences. According to its webpage blurb,

> One of the main goals of Nano2Life is to promote the creation of an RTD-intensive European nanobiotech-related industry. The translation of science into economic benefits is fostered by facilitating the uptake of technology from cutting-edge academic institutions to key sectors such as health, environment, and security.
>
> (Nano2Life, 2008)

Established with the stated role of knowledge transfer and information exchange, the Network seeks to integrate research groups and resources to 'add significant value to European nanobiotechnology research' and to 'speed up the progress of the field' (Nano2Life, 2008).

Investment in fields such as bionanoscience, which some are convinced will 'become one of the key scientific fields of the 21st century', is growing as a consequence of belief in the potential medical and other benefits (Tudelft, 2008). In health and medicine, improved cost savings are expected through the development of more effective drug delivery, enhanced medical imaging and superior implants and prosthetics (RS/RAE, 2004, 20–23). In the fields of agriculture and environment, nanotechnologies are expected to lead to more energy-efficient products and sustainable processes. Nanotechnologies are predicted to contribute to the creation of the zero-carbon, zero-waste society, such as Abu Dhabi's $15 billion Masdar Initiative, which avowedly provides a model for sustainable cities in the future. Funded by IBM and Saudi Arabia's national research and development organization, a new 'green nanotech' lab aims to develop new technologies in solar power, seawater desalination and recyclable materials. Evidently, it represents 'one of several indicators that oil-rich Middle East nations are moving rapidly into clean tech' (LaMonica, 2008).

These 'good news' stories about nanotechnologies reflect a vision of the 'good society', through increased economic well-being, enhanced physical health and a cleaner, more hospitable environment. They also convey belief in the powers and essential beneficence of science: research and development is seen as 'rational' and progressive, the implicit assumption being that a failure to advance science or opposition to developments is somehow 'irrational' and regressive. Bruno Latour has described how the apparent rationality of science serves to exclude that which is seen as 'irrational', including non-scientists' opinions and beliefs about science and its consequences (1987, 180ff). In a society dominated by scientific rationality, it is easy to dismiss 'non-rational' ways of knowing, including lay actors' definitions of risk.

Risks in this context are defined as being largely amenable to rational understanding and management through regulatory measures, as well as by strict adherence to agreed protocols or 'codes of conduct'. In expert discussions, there is rarely acknowledgement of the contending conceptions of risk, nor the significance of the media in framing risk issues (see Chapter 3). Increasingly, scientists are casting themselves as 'responsible' citizens, people who should be trusted to develop technologies in ways that work to the benefit of society and that do not have deleterious or 'risky' consequences.

In recent years, science groups and policymakers in a number of countries have emphasised the importance of the 'responsible' uses of nanotechnologies, which is underpinned by a growing awareness of developments that are likely to impact on communities (e.g. Department of Innovation, Industry, Science and Research, 2008;

RS/RAE, 2004). Reflecting this emphasis in policy, the European Commission announced its adoption of the Code of Conduct for Responsible Nanosciences and Nanotechnologies Research in February 2008. This document acknowledges that there were 'knowledge gaps' in relation to the health and environmental impacts of these technologies as well as issues relating to ethics and respect for human rights. Accordingly, it was recommended that Member States adopt a Code of Conduct governing research in this field. This Code articulates seven principles, which include respect for 'fundamental rights'; concern for 'the interest of the well-being of individuals and society in their [the technologies'] design, implementation, dissemination and use'; and 'openness to all stakeholders' as well as 'transparency and respect for the legitimate right of access to information' (Europa, 2008). However well-intended this language of rights and responsibilities happens to be, it nevertheless risks obscuring the power relations underlying the discursive formulation of problems. That is to say, it fails to address the ways in which certain issues are identified at the expense of those which never become a subject of formal deliberation, namely because they are not defined by key decision-makers as problems in need of action (Bachrach and Baratz, 1962).

The framing of nanotechnologies in terms of 'benefits and risks'

'Risk' is seen as amenable to objective assessment. In its contemporary meaning, 'risk' is calculable and subject to rational management (Castel, 1991). As we noted in Chapter 3, the rationalist approach to risk assumes that, given enough information, it is possible to develop an absolute measure of the riskiness associated with technological innovations. In most contemporary societies, risk is seen as controllable through regulatory measures, such as the improved labelling of consumer products or health and safety standards. In science and policy reports on nanotechnologies, as in the media (see Chapter 3), discussion tends to overlook the uncertainties associated with the field. Questions need to be posed about who ultimately owns (or whose interests are advanced by) particular innovations, who has the right to decide which fields of research and development deserve investment and who determines whether the 'benefits' (however substantial) of particular technologies are worth the attendant 'risks' (however defined). The criteria by which such judgements are determined, it follows, need to be subjected to close inspection.

Scientists, as we have seen in previous chapters, are in a far better strategic position than other actors, particularly lay publics, to impose their definitions of the significance of science and technology developments. Despite their sometimes reductionist depictions of science mediation, they recognise that they are well placed to shape the terms on which information is placed into public circulation. In their communications with stakeholders, as well as with journalists, they are likely to draw selectively on this information to convey a preferred image of nanotechnologies and their applications. In a context of intense competition for research funding and pressures to achieve the next 'breakthrough', scientists are much more inclined to emphasise the value of their research than its attendant uncertainties or dangers.

A study of scientists' and science policymakers' representations of nanotechnologies

That scientists and science policymakers are inclined to emphasise the benefits over the uncertainties and risks of nanotechnologies is confirmed by our study, to which we now turn. As noted, this study explored scientists' and science policymakers' representations of the benefits and risks of nanotechnologies as applied to health and medicine and environmental sustainability, respectively, through the use of online questionnaire surveys and in-depth interviews with a sub-sample of survey respondents who expressed a willingness (on the questionnaire) to be interviewed. The survey employed a combination of fixed choice questions (using Likert scales) and open-ended questions, while the interviews employed a series of semi-structured questions tailored to the individual respondent in order to encourage exploration of their questionnaire responses. The findings from this study highlight scientists' and policymakers' overall optimism about the prospects for this field, despite their frequent acknowledgement of the uncertainties and risks of developments. We suggest that such optimism may lead scientists and policymakers to neglect or underplay the dangers and uncertainties of nanotechnologies in their communications on nanotechnologies and thereby limit public knowledge of this field.

The sample

Respondents were contacted at an international conference, 'Environmental Effects of Nanoparticles and Nanomaterials' (London, 18th–19th September 2006), sponsored by a number of UK government authorities and funding bodies (including The Department of Environment, Food and Rural Affairs (DEFRA)), The Environment Agency (EA) and the Natural Environment Research Council (NERC) as well as, and a number of science societies. This event, which was co-organised by a marine biologist at the University of Plymouth and promoted as the first such conference of this kind, provided an ideal opportunity to contact respondents for our study and to gain a 'snapshot' of scientists' and policymakers' views on nanotechnologies during a period of their growing visibility.[2]

A paper version of the survey was distributed at the conference and non-attendees, whose names and contact details appeared in the conference programme, were sent an email questionnaire. In total, 75 conference delegates were approached in connection with the survey, which elicited responses from 39 attendees, representing a response rate of 52 per cent. A delegate from the above conference subsequently offered to distribute an electronic version of the questionnaire among members of his professional network, the Chemistry Innovation Knowledge Transfer Network (CIKTN, a UK-based knowledge transfer network), which elicited a further 30 completed questionnaires. Finally, an email questionnaire was later sent to a researcher/analyst from an advocacy group involved in public engagement activities who agreed to participate, resulting in 70 completed email questionnaires in total. Twenty individuals who completed the questionnaire survey were then interviewed (one hour average), in order to explore their responses to the survey questions in more depth. Of those who responded at the conference, 35 (90 per cent) were scientists and the rest (4 or 10 per cent) were funders of nanotechnology research. Of those who were members of CIKTN, 26 (86 per cent) were scientists and two (7 per cent) were funders.

We assumed that scientists' and policymakers' views on nanotechnologies, including their benefits and risks, would likely vary to some degree according to their position in the field. Neither constitutes a homogenous group and their perceptions of nanotechnology issues are bound to be shaped by such factors as the research culture in which they work, the length of time they have undertaken the research or been exposed to nanotechnology issues and personal experiences of using nanotechnology-related products. Their role in relation to the production, consumption and assessment of the technology can be presumed to be particularly important in this respect. Consequently, using a typology employed by Powell, the scientists who agreed to participate in our study were classified broadly as either 'upstream' or 'downstream', according to their 'standpoint' in relation to the development of nanomaterials – broadly, designers or developers of new products, or as researchers who research and/or monitor the impacts of developments, respectively (2007, 175).[3]

Our use of 'upstream' in this context, it should be noted, differs from that used in recent discussions about public engagement, which refers to ascertaining publics' views about technologies before significant research and development decisions are made (see Chapter 5). Given the nature of the conference, it is hardly surprising that a large proportion of the respondents from this source (28 of 35; i.e. 80 per cent) were classified as 'downstream' scientists. Of the respondents from the CIKTN group, on the other hand, most (23 of 24; i.e. 96 per cent) worked 'upstream'. The remaining respondents occupied roles that involved a combination of 'upstream' and 'downstream' activities (five from the conference; three from CIKTN) or could not be classified (1) because details of their role were missing. Regardless of their role, the majority of respondents were relatively new to the field of nanotechnologies. Most were found to have only five years or less experience working in the field (42 of 70; i.e. 60 per cent), with only a small proportion (7) with more than 15 years of experience in the field. A slightly lower proportion of 'upstream' scientists (13 of 24; i.e. 54 per cent) than 'downstream' scientists (18 of 28; i.e. 64 per cent) claimed to have had five or less years of experience working with nanotechnologies (eight individuals fell into both categories, while four were unable to be categorised.). Further, a large proportion of the sample worked within Universities (21), the private sector (19) and government (15). Others were employed by a research council (8) or other contexts (6). Finally, 73 per cent (51 of the 70) of respondents were based in Britain and 27 per cent (19) were based overseas (1 in Australia, 1 in Austria, 6 in Denmark, 1 in Germany, 1 in Ireland, 2 in Italy, 2 in the Netherlands, 1 in Sweden and 4 in the USA). Of the 19 respondents based overseas, 79 per cent (15) were conference delegates and 21 per cent (4) were in the CIKTN respondent group.

High expectations

Reflecting the general optimism of the field, noted above, the scientists and science policymakers whom we surveyed and interviewed, regardless of whether they worked 'upstream' or 'downstream', were found to have high expectations about nanotechnologies and saw many applications and benefits in the years ahead. Most survey respondents anticipated that new developments would eventuate within ten years. In the area of medicine and public health, 28 (50 percent) of the respondents saw applications as likely to occur within five years, while 20 (36 percent) thought that they would occur within six to ten years.

Given their frequent acknowledgement of the limits of their own knowledge of nanotechnologies and of the uncertainties of the attendant risks of this field, the basis for this optimism seems to rest largely on faith; a belief in the power and beneficence of science. Many of the respondents acknowledged the difficulty of keeping up with the proliferating literature on nanotechnologies and their reliance on the views of researchers in the field, or what they had read in news media or other sources for their knowledge of developments. Personal experience of nanotechnology products as a consumer and/or family member occasionally influenced personal assessments of the field. One scientist began his interview by explaining that his response to the survey

> was a mixture of things there because some of them were personal and the sort of questions I would expect any reasonably informed person to ask because I'm going to be using and my family, who are very close to me and dear to my heart of course, will be using . . . so there's that sort of general interest.
>
> (Fish and shellfish pathologist and head of research centre)

This respondent's reference to his expectations for a 'reasonably informed person' underlines the difficulties of separating the personal from the professional in making judgements about nanotechnology's implications.

Scientists, of course, do not stand outside of society. They will always bring their personal values to their work, which are likely to influence views on technology developments and decisions about what they study, how they approach problems and how they communicate information. Like other members of society, in their professional and personal lives they are also subject to pressures of time and 'information overload' and are reliant on experts for insights into issues outside their specialist area. Lack of time was frequently cited as a factor in not knowing more about developments, or simply an inability to keep up with the vastly accumulating literature in the field. However, regardless of where they stood in the science–technology cycle, as the surveys and interviews revealed, the respondents in general held a positive view of nanotechnologies and identified a range of potential applications and benefits in the fields of health and environmental sustainability. In the main (subject to qualifications discussed below), they considered the benefits of the applications of nanotechnologies to be outweighing the risks. This often seemed to be a matter of faith, rather than based upon certain objective information about technologies and their development.

As we have noted, many science reports and news items lend the impression that nanotechnologies are revolutionary and that they are set to transform medicine and healthcare, environmental management, materials, information and communications and so on, in the years ahead. In Chapter 5 we referred to the predominance of the more radical perspectives on nanotechnologies in public discourses, which tend to downplay, if not deny altogether, attendant uncertainties. Wood et al. (2008) suggest that developments are likely to be more incremental than is implied by these radical visions. A number of our respondents also questioned the novelty or innovatory aspects of nanotechnologies. As some pointed out, research had been undertaken in the field for many years, if not decades, with some commenting that this work had only recently been 're-packaged', as 'nanotechnologies': for example, 'In my field we have been operating at the nanoscale for decades, so many of the problems and benefits are well known' (research associate, private sector); 'I feel that nanotechnology is just new terminology for science and technology that has been occuring for several years/generations. We are just now in a position to measure it!' (Economic policy manager, development agency).

Like the latter respondent, another pointed out that the key change was the ability to measure at the nano-scale:

> I think it's [nanotechnology] perceived to be a new area; I think it's perceived to be something new and novel and different and, actually, in many cases, its just that the measurement and capability to measure to that scale is now available, so we can measure to that scale and smaller and so we've recognised it and we've given it a classification.
>
> (Economic policy manager, regional development agency)

As some pointed out in the interviews, nanotechnologies constitute a broad field, and different applications will develop at an uneven pace:

> I would say that ... the whole thing about nano is it's different generations of ... a vast range of technologies, it's not just one thing and there's going to be different aspects of that coming through over the next forty or fifty years. I don't think it's going to be a giant leap forward as much as an incremental ... sort of growing within various sectors.... I do think they will start to come through ... over the next five years.
>
> (Chief scientist, government agency)

The view of nanotechnologies as non-novel and incremental presented here stands in contrast to the much more prevalent public representations typically promulgated through news media and other forums. As discussed earlier, journalists tend to focus on issues that are judged to be 'newsworthy' – the spectacular, dramatic or unexpected – rather than on mundane happenings. Scientists themselves appear to be well aware that the incrementalist view on science development is unlikely to attract news media interest, nor the attention of funders of research (who, it is assumed, are more likely to be persuaded in their decisions by research which is seen as innovative and likely to lead to applications within the short-to-medium term). There would seem to be an imperative then to 'hype' nanotechnology innovations by reference to the more radical visions.

Views on risks and their assessment

In the survey, respondents were asked whether they perceived specific risks associated with nanotechnologies, and if so whether there were any particular applications that were 'especially risky'. They were also asked to rate the levels of risk they saw associated with the applications of nanotechnologies in the fields of medicine and public health and environmental sustainability, respectively, as well as their knowledge of the risks in these fields. A Likert scale was employed in each case. In interviews these responses were used for exploring their constructions of risk in more depth. After reminding them of how they rated the items in the surveys, we asked them, 'do you still agree with your assessment?' Moreover, we invited them to elaborate, with questions such as 'how did you come to learn about those risks?'

In the surveys, all but one respondent indicated that they saw risks associated with nanotechnologies. This dissenting individual, a regulatory toxicologist and consultant, added a note to his survey: 'We need to evaluate then estimate exposures to assess risks. Do not believe we have sufficient hazard data yet and also we should not leap into "grouping" of nanoparticles until we have adequate, robust data based on well-characterised materials.' When asked to state the risks, the responses were varied, but tended to cluster around references to dangers to humans (e.g. 'skin and lung penetration') and environments, and public responses (e.g. 'public acceptability', 'misunderstanding of science by the general public'). When these responses were probed in interviews, however, a more complex picture emerged.

'A lot of unknowns'

Many respondents emphasised the difficulty of assessing risk due to the lack of research and knowledge about the pertinent issues. Respondents pointed to difficulties in characterising particles or their behaviour in certain environments, lack of investment in risk assessment, uncertainty about the adequacy of established methods for ascertaining nanoparticle risk and what one described as the 'latency period' between the introduction of technologies and the evaluation of their potential adverse effects. A typical response, articulated by a scientist, who works with government agencies formulating nanotechnology policy, is that 'it is really hard for me to gauge [the risks of nanotechnologies] because it's just a lot of unknowns'. As the above respondent observed, when commenting on environmental risks of nanotechnologies, 'I guess there are risks with any manu-facture of nanoparticles, there's no real guarantee that they're being kept in the system and there's no real tests to detect where they are and what damage they're doing.' The comments of one respondent, namely that the newness of the field made risk assessment difficult, were not uncommon: 'I mean there's not as if there's like a massive history of research or anything or work that's been done that is used as a basis for risks, because they're new particles...they have to be tested' (Science programme officer, research council).

Some pointed to the difficulty of assessing risk when the applica-tions are uncertain. In other words, as would be expected, perceptions of risk reflected respondents' locations within the science produc-tion/consumption/assessment cycle. Those whose work focuses on risk assessment can be expected to have especially strong views on the nature and challenges of establishing risk. When asked to explain their 'high' rating of the risks of nanotechnology applications in the area of environmental sustainability in the survey, a science policy advisor, who specialises in hazard/risk assessment policy develop-ment, emphasised the challenges of assessing risk. Like a number of others, he changed his assessment of risk, in this case to 'neutral', explaining:

> I think it's quite dangerous to speculate is what I'm saying. I mean, just because...you know, you've got a nanoparticle doesn't nec-essarily mean that it'll pose a hazard, because it's all related to exposure. So unless people are going to be exposed or the envi-ronment seriously exposed.... Some of these applications, if they are in batteries for example...they are not actually going to be used directly in the environment, it's just that they will have an environmental benefit associated with them and it's difficult to see what the exposure might be to the nano, um, dimension of those batteries. It might be part of the battery itself and so there might be an issue of when the batteries are disposed, for exam-ple, or whether they're going into the environment. But that is an extremely difficult thing to comment on, on what the level of risk might be in the absence of knowing, you know, the hazard of those particular nanoparticles.
>
> (Senior Policy Advisor, hazard/risk assessment
> policy development)

Another respondent, who specialised in fish and shellfish pathology, thought that the risks posed by nanoparticles were not high, but he did emphasise the lack of assessment of risks. As he explained,

> From the knowledge that I have…a lot of the substances or the nanoparticles that have been used are actually bound within other materials for us and…in that context…I would have thought that the possibility for them to be released in any quantity to cause harm in the environment would be relatively low…so that's what I think, that there are risks but they're not properly assessed yet I don't think.
>
> (Fish and shellfish pathologist and head of research centre)

Especially risky?

In the study, we asked respondents whether they thought there were applications that were 'especially risky' and, if so, to identify these risks. Of the 60 who answered this question, 52 (87 per cent) answered 'yes', and among the scientists who responded (47) an approximately equal number worked 'upstream' (21) and 'downstream' (19) (the other six were classified as 'upstream and downstream'). Given the backgrounds and interests of the respondents in the sample it is perhaps not surprising that the responses tended to cluster around the risks of inhalation and ingestion of 'free nanoparticles' arising from the manufacture and use of nanomaterials/products (sunscreens, food, cosmetics, medical implants, fuel additives, aerosols and remediation of ground water were mentioned). In interviews, many respondents emphasised that medical applications had a 'high' risk or were 'especially risky'. One respondent reflected the views of a number of others in commenting: 'I think there are risks with all sort of new medical interventions, so, in a sense, those in nanotechnology are a sort of subset of that…both in terms of whether there are immediate side effects or any longer term issues associated with them' (Funder, research council).

One respondent reasoned, 'I think the risks [associated with the application of nanotechnologies in medicine and public health] are a little higher than average because…you're putting things into people's bodies, so there's always going to be a slightly higher risk.' This respondent then went on to explain that they believed that there were likely to be 'toxicological effects' or 'bioaccumulation of active ingredients in the body which may not have happened at the kind of larger particle size' (Technology platform manager, knowledge transfer).

Respondents frequently identified the potential adverse effects of accumulation of nanoparticles in the body and the environment as a problem. A risk was sometimes defined as 'high' when there was seen to be uncertainty about the longer term effects of nanotechnologies, which are seen to have unique properties:

> I think the high risk comes from the fact that if you are talking about health or putting things in people's bodies...until sort of ten, fifteen, twenty years down the line you don't really know what the effect will be. You can do clinical trials but...until you've actually got the sample size big enough you don't know what the true benefit or, indeed risk, could be. I think health has a particularly high risk with it anyway. So if we use something that has got unique properties because it is a nanoparticle then it could also have unique...disadvantages or side effects that perhaps you don't see in the first instance. But then I think there are, you know, lots and lots of low risk applications...creams and things like that that are in general quite a low risk.
>
> (Economic policy manager, regional development agency)

Another respondent explained that there were 'always risks' that 'you've got to address at some point in the...development of your product'; however,

> one of the risks that always comes home to me is the use of enzymes in...some of these products. I mean the glucose biosensor, and some of the cardiac risk ones that I'm...developing, use enzymes.... these are fine outside of the body, perfectly fine, but I wouldn't want to get an enzyme loose inside the body that isn't a natural one because it'll produce...something like toxic shock syndrome, inflammation, possibly death. Uh, so, you have to be very, very careful about introducing free nanoparticles, proteins, or what have you, into the body. You've got to have done the tests first to make sure they're not going to do damage to the person.
>
> (Scientist and director, science research centre)

One respondent, a science programme officer at a research council, on the other hand, saw the environmental risks as being higher than the medical applications:

> Well, there's obviously... a much bigger exposure route for a large number of people if you're using it [nanotechnology] in environmental situations because... if you are actually giving it to somebody as an individual then they're... the people that are getting exposed, whereas if you're putting it in to land or you're getting into waterways and things like that, then obviously that gets recycled through waste and water treatment works and things like that, so potentially you could expose a lot more people to it but probably at a lower concentration.
>
> (Science programme officer, research council)

Views on benefits versus risks: Blind faith?

When asked in the interviews whether they believed the benefits associated with the applications of nanotechnologies outweighed the risks, some respondents pointed to the difficulty of making such an assessment in the absence of information about specific applications and clearer information about both the benefits and the risks. As one respondent (Scientist, government agency) commented, 'the risks and benefits have not been sufficiently characterised and communicated for us to be able to make the judgement'. In the vast majority of cases, however, when asked, respondents said that they saw the benefits of the technologies as 'outweighing' the risks. As noted, despite the acknowledged 'unknowns' about the risks, most provided a positive assessment of nanotechnologies and anticipated imminent beneficial application in the fields of health and medicine and environmental sustainability in the foreseeable future. The comments revealed that the majority subscribed to a progressive view of science and, mostly, faith that the regulatory systems can control the risks *once they are assessed*. For example:

> The benefits outweigh the risks once you've taken into account the regulatory frameworks that will exist around most of the applications.
>
> (Funder, research council)

> I think the potential benefits outweigh the potential risks. I
> think...my view is that...everything we do you have to manage
> the risk and if you can manage the risk successfully then I think
> that the benefits outweigh the risks.
>
> (Scientist, working with research councils)

> The benefits will outweigh the risks because...in a situation
> where governments are aware that this is an emerging technol-
> ogy...everything is in place to be able to...with accumulated
> knowledge assess risk in the appropriate way. And so I think that
> any risks should be...well managed and...that the benefits will
> definitely outweigh the risks.
>
> (Fish and shellfish pathologist and head of research centre)

As one respondent emphasised, the risks of applying nanotechnologies to solve environmental problems was 'lower than doing
nothing...because there are so many areas that are polluted very,
very badly' (Scientist and director, science research centre). He went
on: 'it does strike me that when we worry about the risks of nanoparticles which are engineered by me and my colleagues now we worry
more about that then we do about the car tyre which is releasing
billions of nanoparticles per revolution of wheel every day'. Another
respondent argued,

> I just think there's a risk in anything, any new area or any new,
> not necessarily new, but any kind of technology there's a risk asso-
> ciated with it and the key to minimise the risk is understanding
> that. I think that we are now at a stage where we can measure,
> characterise, clarify what we're actually using.
>
> (Economic policy manager, regional development agency)

Another respondent, in explaining why he rated the risks associated
with applications in the field of medicine and public health as 'low',
explained,

> Well the reason I say that is is that...before any nanotechnol-
> ogy development in that area becomes something which is used it
> would obviously go through a very comprehensive...risk assess-
> ment process and therefore...it wouldn't be on the market or

allowed to be used if the risks were high. So ... the answer is that you'll only ever have products in the area that are low risk or, otherwise they won't go through the system.

(Science policy advisor, hazard/risk assessment
policy development)

Public responses to nanotechnologies

This overall optimism in relation to nanotechnologies was tempered, in many cases, by concerns about public acceptability or the impact of public responses to nanotechnologies. As noted, some respondents saw public reactions as a 'risk' in that they could potentially adversely affect developments. A number made explicit references to the GM controversy of the late 1990s. For example:

I'm very optimistic, I think, that this is a technology which is going places. I think there's a lot of concerted effort around the world to make sure that the implication work keeps up with the application work and that, you know, not too many ... things happen, or products are developed and put into the market place without due consideration of the implications ... We've just got to hope that, you know, something doesn't happen, an incident, you know, to turn public opinion against ... the technology and generally brand the area as ... something similar to GM.

(Science policy advisor, hazard/risk assessment
policy development)

Public response to applications of nanotechnology in food production was a focus of particular concern:

The more I learn about potential uses of nanotechnology in foods, the more I'm concerned about the fact that this could go in the same way as the GM debate went and it could actually be a trigger that creates a lot of public feeling against nanotechnology if it's not regulated and looked at properly.

(Principal scientist, research institute)

I don't know of any but if somebody was to say to me that they were going to use nanomaterials in some way or another with

food production, I think that would be a natural one for extreme
caution and public anxiety probably.

(Fish and shellfish pathologist and Head of research centre)

These concerns are not unfounded. Some consumer action groups
and environment groups have voiced their alarm about nanomaterials in food and food-storage products. For example, FOEA has
warned of the dangers of 'nanofood' and challenged the notion that
nano-enhanced products necessarily benefit publics, especially those
people in the less-affluent groups who are least able to afford such
products. It has called for a moratorium on the use of nanotechnology for the food sector and the regulation and labelling of food,
food packaging and agricultural products that contain manufactured
nanoparticles, before allowing any further commercial sales (FOEA,
2008a, 47–48).

Among our respondents, there was acknowledgement of the problems of labelling products as 'nano' to consumers. Increasingly
environmentally conscious publics may be concerned, for example,
about what happens to products labelled 'nano' once they are thrown
away. As one respondent said, 'how it's currently described will result
in whether we see a lot of these technologies take off or whether
they're restricted because of ignorance or misunderstanding or fear'
(Economic policy manager, regional development agency). Further,
some industries were 'suspicious' of nanotechnologies, as a scientist
involved in technology transfer explained,

> if you come along now with a new company with 'nano' in the
> name, either in the prefix or suffix of it, venture capital people and
> the people in the know in the big corporations really get suspicious
> and start asking all sorts of questions, and it becomes a bit of a
> turn off.
>
> (Scientist and Head of research centre)

Conclusion

Scientists and policymakers who seek to promote nanotechnologies
are confronted with a dilemma in their communication efforts. In
order to engender support for innovations they are liable to 'hype'

the benefits and/or downplay the negative implications, including the risks to health, safety and environment. However, like other emergent technologies, there are many uncertainties in relation to the trajectory and impacts of developments. The challenge would seem to be especially acute with nanotechnologies given the breadth of potential applications and the lack of a widely agreed language, metaphors and analogies for characterising the field. As we noted in earlier chapters, this is a field where both experts and popular writers struggle to find meaningful imagery. In the event this imagery is unclear, and/or if the risks are uncertain or perceived to outweigh the benefits, then it is more than likely that technologies will be resisted, if not rejected outright. Hence, the question of how scientists and science policymakers represent nanotechnologies is likely to be crucial to future publics' perceptions of the field.

The shift in rhetoric in science communication from 'public understanding' to 'public engagement', we contend, has given rise to an implicit tension in the society–society relationship. 'Engagement' or 'dialogue' with 'the public', at least as it is routinely articulated in policy documents in the UK, Europe and Australia, suggests that scientists and policymakers will seek to communicate information of a kind and in a manner that will allow 'the public' to fully understand the nature, applications and implications of nanotechnologies in order that they may make 'informed' decisions. This presupposes some level of agreement among scientists and the various stakeholders about what constitutes a 'benefit' and a 'risk' and how they should be assessed. It similarly suggests that a sufficient degree of 'openness' or 'transparency' will be assured in relation to the information that is conveyed. However, as we have argued, and as our data confirm, scientists and policymakers do not hold a singular view on nanotechnologies and their risks. The novelty of nanotechnologies is questioned by some experts and there are seen to be many 'unknowns' about the risks and the adequacy of measures of risk assessment. Nevertheless, they are broadly in agreement that benefits will accrue from innovations and that these will outweigh the risks, once developments are appropriately regulated. This confidence would seem to be a matter of faith rather than based on empirical evidence. Our respondents' comments, on the whole, reflect a belief in the beneficence of science and in the regulatory institutions that govern resulting technologies. Still, it could be argued that, given the acknowledged gaps in the understanding of risks, the basis for this optimism is far from firm.

Uncertainty in science communication presents a difficult challenge, yet one which needs to be met by both scientists and regulators alike. Some questions which arise are: What role should regulation play in this context of uncertainty? In particular, what exactly is it that is to be regulated and with what objective? Is there something particular about nanotechnologies that call for specific regulations – in relation to enabling developments and managing risks? What do publics – the potential users of nanotechnologies – need to know about the technologies' attendant risks, including threats to health and well-being and the environment? Science is unable to provide clear guidance on these questions, since they are matters for public deliberation rather than scientific determination, necessitating wide-ranging discussion and input from the various publics and stakeholder communities. As the history of regulatory efforts (e.g. in public health) show, regulation can have unanticipated effects and, when unsupported, can engender resistance and fail. An open and thorough debate about the above questions in the light of the 'unknowns' about nanotechnologies can help provide a basis for regulations which not only protect heath and environment but engender trust among publics that is necessary to ensure the realisation of useful technologies in the future.

Notes

1. Specifically, most of the data for this chapter is drawn from a British Academy-funded study undertaken by two of this book's authors, Petersen and Anderson. For further details, see Petersen and Anderson (2007).
2. In the domain of science and technology, 'policymaker' may be conceived broadly to refer to formal political actors (politicians) who make decisions that shape policies affecting the overall direction and practices of science and technology at the international, national, regional and local levels, as well as local decision-makers who play a formal or informal role in the multitude of day-to-day decisions about the funding, conduct and assessment of research and development; for example, research council programme officers, research centre managers, members of research ethics committees, science policy advisors, technology transfer specialists. Scientists often draw a distinction between science/scientific research, on the one hand, and society/policy-making, on the other, which is far from clear

in practice. The drawing of such a distinction may serve in 'boundary work' or the rhetorical purpose of buttressing the authority of science and its claim to be untainted by social influence or interests and thus objective or value-free (see, e.g. Gieryn, 1999).

3. As Powell explains, '"Upstream" scientists design and develop new (and usually synthetic) materials', and include engineers, chemists, physicists, materials scientists and, increasingly, biologists (2007, 175). In her view, they are concerned with the properties and characteristics that make materials work. 'Downstream' scientists, on the other hand, comprise 'toxicologists, epidemiologists, and other public health scientists' who study and monitor the environmental materials that are created by the 'upstream' scientists (2007, 175).

7
Communication about Nanotechnologies in the Future

The previous chapters examined how nanotechnologies have been framed in the media, by scientists and within the broader policy arena. Given the currently low but growing level of public visibility of nanotechnologies, we believe that it is timely to reflect on how information about these technologies is currently being communicated and also try to look ahead to the future.

In the light of current predictions that nanotechnologies will find an array of new applications in the coming years, questions about the ways in which facts – but also values – are communicated during this emergent phase are likely to prove crucial at a number of different levels. How publics will respond to this field is certain to be influenced by the informational resources available to them, for example, which will in turn affect how particular innovations develop over time. As we have noted in earlier chapters, there has been a remarkable array of contrary claims about the potential of nanotechnologies to transform society, involving both utopian and dystopian scenarios. Our analysis of UK print news media during this early phase (Chapter 4) revealed some of the diversity of these portrayals. Nanotechnologies are frequently characterised by their novelty. However, we have also seen how the term 'nanotechnology' itself is mired in definitional ambiguity, a problem compounded by various competing representations of the benefits and risks of particular innovations. As a number of scientists whom we interviewed indicated, this is a field with many 'unknowns' (Chapter 6). The UK's RS/RAE have also acknowledged the uncertainties of nanotechnologies, especially in relation to the risks posed by manufactured nanoparticles (RS/RAE,

2004). In this context, it needs to be asked, how should publics respond to these portrayals? Specifically, how may they adequately appraise the often conflicting claims about the benefits and risks of nanotechnologies? What kinds of information about nanotechnologies needs to be conveyed and how should this be presented in order to allow publics to properly assess their impacts?

Science communication efforts in this area, we have stressed at various points in this discussion, have been recurrently based on a number of enduring and, we contend, questionable assumptions about the science–society relationship, the processes of communication, and 'the public'. Science tends to be presented as somehow outside society, as constituting unmediated truth, with 'communication' being conceived as a process of informing an assumed 'ignorant' or 'unaware' audience about 'the science facts'. This simplistic, linear model of science communication is premised upon a particular view of the rationality of public action. A more enlightened or 'aware' public, technology proponents assume, will be receptive to technologies and perhaps less likely to reject potentially valuable applications. Recent 'public engagement' efforts in relation to nanotechnologies, although presented as enlightened efforts to democratise science, tend to be mainly variations of the so-called 'deficit' model of the public understanding of science. That is, implicitly, they assume that 'the public' is ignorant or unaware of science, and therefore needs to be educated through more or 'better' information, with scientists positioned as the arbiters of 'truth'. There is rarely a proper acknowledgement of the existence of diverse publics and stakeholders, with different perspectives on the field and investment in the issues, whose lives will be unequally affected by developments. Scientists constitute but one of these groups of stakeholders, albeit an especially powerful one. Public engagement, we observed, has tended to operate in practice as a kind of risk management strategy, a means of pre-empting and defusing opposition, rather than encouraging wide-ranging debate about and deliberation upon the substantive issues raised by nanotechnology innovations. This approach leaves unexamined expert knowledge itself and how it is represented and conveyed in the news and other forums.

Drawing upon our own empirical research in this field, we have argued for a more theoretically sophisticated understanding of the science–society relationship. Our starting point is the recognition of the contemporary socio-political significance of the media, in framing issues and establishing the agenda for public debate and action. The tendency of scientists to treat the media as a 'blackbox' and to overlook their own role as key expert sources in news stories, noted in Chapter 5, can itself be seen as an aspect of the playing out of media politics. By presenting themselves as victims of media distortion, some scientists have sought to define and police the boundaries between 'real' and 'distorted' science, thus buttressing their own authority (Gieryn, 1999). This is not to deny the particular imperatives that shape the media reporting of issues. However, a more critical, reflexive approach to the communication of science would entail interrogating processes of science communication (the 'blackboxed' inner workings of the media), not least by recognising scientists' own role as significant claims makers and by exposing their assumptions, values and 'deficits' of understanding about 'the public'. This line of investigation is necessary if we are to narrow the gap between lay and expert understandings of technologies, surely a precondition for democratic public deliberation on the pertinent issues. In short, 'science communication' itself, we believe, needs to be unpacked and rethought.

If the science–society relationship is to be reconfigured it will need to be acknowledged at the outset that science and technology are both inescapably *mediated*. Further, the mediation process is an inherently political one; that is, it involves competition between competing claims makers who seek to present their particular 'spin' on issues. This means that only some stories about science and technology get told – those of the stakeholders who are successful in helping to set and reinforce the news agenda. The concept of framing, we argued, is useful in this context, drawing attention to the selectivity involved in the portrayal of information. How publics understand nanotechnologies (or indeed any technologies) and their implications is likely to depend crucially on how information is framed in the news media and other forums. In a media-saturated society, publics rely more than ever before on newspapers, the Internet, the television and radio for information about the nature, benefits and risks of new technology innovations.

The potential of the media to profoundly affect our view of the world has been noted by scholars at least as far back as the earliest days of the printing press. Today, the growing convergence of technologies facilitated through the electronic revolution has vastly accelerated the pace of information flow such that it is difficult for most people to find a point of reference outside the institutions of media from which to assess the competing and constantly circulating claims about science and technology. Consequently, in order to be able to properly assess the competing claims and portrayals of nanotechnologies it is important for publics, which include scientists, to develop new forms of media engagement. To consider what this may entail, we have examined some recent evidence from the literature on science and the media over previous chapters. Singled out for attention has been those factors that may affect how news stories about science are framed and how technological risks are constructed (Chapters 2 and 3). As we noted, this work highlights how the day-to-day routines of journalistic practice may influence what gets in the news and how it is represented. Our own work highlights that certain framings of nanotechnologies are shaped by the priorities of news reporting, such as the concern with celebrity figures (Prince Charles' intervention in 2003) and the focus on the controversial and more radical visions of innovations; for example the so-called 'grey-goo' scenario (Chapter 4). We noted that during the emergent period of nanotechnology's arrival on the news agenda, the ensuing press coverage was generally positive in tone. However, different sections of the press provided varying degrees of attention to technologies, offering surprisingly diverse inflections to issues. Stories about the substantive social, economic and political implications of innovations were relatively scarce across all sections of the press. Although we are unable to assess the impact of this period of coverage on publics' views, we would argue that it is likely to have worked against a broad public appreciation of the nature and diverse implications of technologies. Such coverage certainly does not encourage *sustained* wide-ranging discussion and deliberation on the issues.

Our project was limited to a period in which nanotechnologies received heightened attention in the UK press, partly as a consequence of a number of prominent nanotechnology-related issues under discussion at the time. The project did not trace the rise and fall of attention to nanotechnology issues over an extended period of time or compare coverage of these with other technology issues such as genetics. Hence, it is difficult to track the relative prominence of coverage given to these technologies and how this may impact on levels of public knowledge. However, by exploring the coverage of nanotechnologies in depth over *all* major UK national newspapers we were able to provide a more nuanced picture of news coverage than would have been presented by focusing on only a section of the press, and/or by tracking coverage in less depth over a greater time span. We observed that the 'elite' press, which is oriented to a relatively restricted readership, gave more focused attention to nanotechnology issues. A number of the scientists who participated in our project acknowledged that different forms of media may present a different picture of issues, with some outlets, particularly the 'elite' media, providing more 'accurate' coverage. We suggested that the limitation of the coverage of issues to the 'elite' press may at least partly account for the apparent low level of public visibility of nanotechnologies noted in some science reports. Future communication efforts in this area need to pay cognisance to where publics are likely to derive their information, and how the type of the media outlet may shape what and how information is represented.

Scientists, it follows, would benefit from a more rigorous understanding of the significance of the rapidly changing character and day-to-day operations of the media for their own work. The growing use of the Internet, particularly among young people, including the use of blogs (see Chapter 2), is profoundly altering the nature and circulation of information (Allan, 2006; Couldry et al., 2007; Hermida, 2006). Sources are more diverse and consequently scientists may have less control than in the past over the framing of issues. It is difficult to predict the impact of a broader range of news sources on the nature of coverage, but the entrance of a wider array of claims makers is likely to broaden the spectrum of competing frames. As coverage of nanotechnology acquires a higher profile within press coverage, the debate is likely to become more polarised, possibly reflecting a shift from predominantly scientific frames to a greater concern with social implications and controversy (see Anderson et al., 2005; Te Kulve, 2006).

Scientists, in our view, need to further reflect on how their own established patterns of working and media relations, including links with preferred news outlets or particular journalists, may shape what and how information is portrayed. Pressures to gain publicity for their work in order to secure grant funding may lead them, in some cases, to hype or 'spin' their own research, thus lending support to the more positive radical visions of nanotechnologies. The collective impact of individual scientists hyping their work may result in a very partial view of nanotechnologies, one which is either overly optimistic or unduly pessimistic. Simply blaming 'the media' or sections of the media for purportedly 'inaccurate' reporting, as did some of the respondents who participated in our research, does not advance debate about what might constitute a scientifically 'accurate' or more adequate portrayal of technologies.

Questions that need to be raised include, how do scientists' particular use of language and imagery translate in the context of the media? What kinds of information are essential for effective communication, and what analogies are appropriate for conveying the complexities of an emergent field such as nanotechnologies? What constitutes 'effective' communication? To what extent do scientists themselves, through their particular use of language, contribute to the generation of controversy – even the climate of fear that sometimes accompanies news stories of technological 'breakthroughs', such as human cloning, or GM food and crops? Scientists tend to portray their work as devoid of metaphorical content; however, metaphors are an inescapable part of science and indeed are essential to the creativity and communication of science (Maason and Weingart, 2000). When seeking to convey complex issues and information about nanotechnologies, they need to find a language that connects with readers and audiences (Petersen, 2005, 204). However, as we have suggested, scientists themselves struggle to agree on an appropriate language and have different views on what is to be communicated and how it is to be conveyed. Those whom we surveyed and interviewed were frequently dissatisfied with many of the prevailing images of nanotechnologies but were less clear about those that would more adequately convey the issues. More thought needs to be given to what are appropriate terms, metaphors or analogies for representing new technologies, and what kinds of images should be used to convey the complexities and uncertainties of this field.

The predicted convergence of a range of technologies in the future – for example, nanotechnologies, information technologies, genetics and the neurosciences – is likely to present challenges for science communication that are as yet unimagined. To the extent it is realised, such convergence will affect how we see and interact at many levels in everyday life. The ubiquitous use of the Internet and mobile phones has already transformed how we communicate and how we see ourselves, something which has occurred without wide-ranging public debate. If publics are to have a role in shaping technologies in the future it is important to develop appropriate means for engaging their participation well before such technologies become firmly established. Every effort must be made to facilitate the broadest input of views into decisions about which innovations are to be developed, how and why.

Clearly, then, such initiatives will require fresh approaches to familiar problems. Enhancing media engagement along the lines briefly sketched above, we believe, has the potential to reinvigorate the democratisation of science and technology, especially where uncertainty proves controversial. Nanotechnology may profoundly shape everyone's lives in ways only now beginning to be anticipated, so it is vitally important that as broad a range of stakeholders and publics as possible have a say in the direction of possible developments. After all, history suggests that it is all too likely that choices about nanotechnology, made in the absence of sustained dialogue, deliberation and debate, will advance particular interests at the expense of those that are demonstrably of value to the population as a whole. In other words, we fear, lost in the rhetoric of 'the public interest' may be a real and meaningful engagement by diverse publics in their future.

References

Adam, B. (2000a) 'The Media Timescapes of BSE News', in S. Allan, B. Adam, U. Beck, and J. Van Loon (eds) *Risk Society and Beyond*: *Critical Issues for Social Theory*. London: Sage.

Adam, B. (2000b) 'The Temporal Gaze: The Challenge for Social Theory in the Context of GM Food', *The British Journal of Sociology*, 51 (1), 125–142.

Adam, B., Beck, U., and J. Van Loon (eds) (2000) *The Risk Society and Beyond*: *Critical Issues for Social Theory*. London: Sage.

Ali, N., Lo, T., Auvache, V., and P. White (2001) 'Bad Press for Doctors: 21 Year Survey of Three National Newspapers', *British Medical Journal*, 323 (7316), 782–783.

Allan, S. (2002) *Media, Risk and Science*. Buckingham and Philadelphia: Open University Press.

Allan, S. (2004) *News Culture* (second edition). Maidenhead and New York: Open University Press.

Allan, S. (2006) *Online News*: *Journalism and the Internet*. Maidenhead and New York: Open University Press.

Allan, S. (2008) 'Making Science Newsworthy: Exploring the Conventions of Science Journalism', in R. Holliman, J. Thomas, S. Smidt, E. Scanlon, and L. Whitelegg (eds) *Investigating Science Communication in the Information Age*: *Implications for Public Engagement and Popular Media*. Oxford: Oxford University Press.

Allan, S., Adam, B., and C. Carter (eds) (2000) *Environmental Risks and the Media*. London: Routledge.

Allan, S., Anderson, A., and A. Petersen (2005) 'Reporting Risk: Science Journalism and the Prospect of Human Cloning', in S. Watson and A. Moran (eds) *Risk, Trust and Uncertainty*. London: Palgrave.

Anderson, A. (1993) The Production of Environmental News: A Study of Source-Media Relations. Unpublished PhD thesis, University of Greenwich.

Anderson, A. (1997) *Media, Culture and the Environment*. London: UCL.

Anderson, A. (2003) 'Environmental Activism and News Sources', in S. Cottle (ed.) *News, Public Relations and Power*. London: Sage.

Anderson, A. (2006) 'The Media and Risk', in G. Mythen and S. Walklate (eds) *Beyond the Risk Society*: *Critical Reflections on Risk and Human Security*. Buckingham: Open University Press.

Anderson, A., Allan, S., Petersen, A., and C. Wilkinson (2008) 'Nanoethics: News Media and the Shaping of Public Agendas', in R. Luppicini and R. Adell (eds) *Handbook of Research on Technoethics*, Information Science Reference.

Anderson, A., Petersen, A., and M. David (2005) 'Communication or Spin? Source-media Relations in Science', in S. Allan (ed.) *Journalism*: *Critical Issues*. Maidenhead and New York: Open University Press.

Anderson, A., Petersen, A., Wilkinson, C., and S. Allan (2005) 'The Framing of Nanotechnologies in the British Newspaper Press', *Science Communication*, 27 (2), 200–220.

Anderson, B. (2007) 'Hope for Nanotechnology: Anticipatory Knowledge and Governance of Affect', *Area*, 19 (2), 156–165.

Australian Nanotechnology Alliance (2007) Australian Nanotechnology Alliance: Supporting a New Era in Traditional Industries, Media Release, 23 November 2007. http://www.nanotechnology.org.au/documents/media_releases/Media_Release-AGM_Nov_2007.pdf accessed 7th June 2008.

Bachrach, P. and M. Baratz (1962) 'The Two Faces of Power', *American Political Science Review*, 57 (December), 947–952.

Bainbridge, W.S. (2002) 'Public Attitudes towards Nanotechnology', *Journal of Nanoparticle Research*, 4 (6), 561–570.

Bainbridge, W.S. (2004) 'Sociocultural Meanings of Nanotechnology: Research Methodologies', *Journal of Nanoparticle Research*, 6 (2), 285–299.

Bartlett, C., Sterne, J., and M. Egger (2002) 'What is Newsworthy? Longitudinal Study of the Reporting of Medical Research in Two British Newspapers', *British Medical Journal*, 325 (7355), 81–84.

Bauer, M.W. and G. Gaskell (eds) (2002) *Biotechnology*: *The Making of a Global Controversy*. Cambridge: Cambridge University Press.

Bauer, M.W. and M. Bucchi (eds) (2007) *Journalism, Science and Society*: *Science Communication between News and Public Relations*. London: Routledge.

Beck, U. (1992a) *Risk Society*: *Towards a New Modernity*. London: Sage.

Beck, U. (1992b) 'From Industrial Society to Risk Society: Questions of Survival, Social Structure and Ecological Enlightenment', *Theory, Culture and Society*, 9 (1), 97–123.

Beck, U. (1996) 'Risk Society and the Provident State', in S. Lash, B. Szerzynski, and B. Wynne (eds) *Risk Environment and Modernity*: *Towards a New Ecology*. London: Sage.

Beck, U. (2002) 'The Silence of Words and Political Dynamics in the World Risk Society', *Logos*, 1.4. http://logosonline.home.igc.org/beck.htm accessed 19 February 2008.

Beck, U. (2006) *The Cosmopolitan Vision*. Cambridge: Polity.

Bell, D. (2005) *Science, Technology and Culture*. Maidenhead and New York: Open University Press.

Bell, T.E. (2006) Reporting Risk Assessment of Nanotechnology: A Reporter's Guide. National Nanotechnology Initiative, http://www.nano.gov/html/news/reporting_risk_assessment_of_nanotechnology.pdf accessed 16 May 2008.

Berube, D. (2006) *Nano-Hype*: *The Truth behind the Nanotechnology Buzz*. Ambherst and New York: Prometheus Books.

Best, S. and D. Kellner (2001) *The Postmodern Adventure*. London and New York: Routledge.

Blum, D. and M. Knudson (1997a) 'Editors' Note', in D. Blum and M. Knudson (eds) *A Field Guide for Science Writers*. New York: Oxford University Press.

Blum, D. and M. Knudson (1997b) 'Techniques of the Trade', in D. Blum and M. Knudson (eds) *A Field Guide for Science Writers*. New York: Oxford University Press.

Bourdieu, P. (1998) *On Television and Journalism*. London: Pluto Press.

Bowman, D.M. and G.A. Hodge (2007) 'Nanotechnology and Public Interest Dialogue: Some International Observations', *Bulletin of Science, Technology and Society*, 27 (2), 118–132.

Broks, P. (2006) *Understanding Popular Science*. Maidenhead and New York: Open University Press.

Brossard, D., Shanahan, J., and T.C. Nesbitt (eds) (2007) *The Public, the Media and Agricultural Biotechnology*. London: CABI.

Brown, N. (2003) 'Hope against Hype – Accountability in Biopasts, Presents and Futures', *Science Studies*, 16 (2), 3–21.

Brown, N. (2006) 'Shifting Tenses: From "Regimes of Truth" to "Regimes of Hope." ' University of York, UK: Science and Technology Studies Unit.

Brown, N. and A. Webster (2004) *New Medical Technologies and Society: Reordering Life*. Cambridge and Malden, MA: Polity:

Bubela, T.M. and T.A. Caulfield (2004) 'Do the Print Media "Hype" Genetic Research? A Comparison of Newspaper Stories and Peer-reviewed Research Papers', *Canadian Medical Association Journal*, 170 (9), 1399–1408.

Bucchi, M. (1998) *Science and the Media: Alternative Routes in Scientific Communication*. London and New York: Routledge.

Burgess, A. (2004) *Cellular Phones, Public Fears and a Culture of Precaution*. New York: Cambridge University Press

Carvalho, A. (2007) 'Ideological Cultures and Media Discourses on Scientific Knowledge: Re-Reading News on Climate Change', *Public Understanding of Science*, 16 (2), 223–243.

Carvalho, A. and J. Burgess (2005) 'Cultural Circuits of Climate Change in UK Broadsheet Newspapers, 1985–2003', *Risk Analysis*, 25 (6), 1457–1469.

Castel, R. (1991) 'From Dangerousness to Risk', in G. Burchell, C. Gordon, and P. Miller (eds) *The Foucault Effect: Studies in Governmentality*. London: Harvester Wheatsheaf.

Chalmers, M. (2007) 'A Revolution in Bits', *Physics World*, 20 (1), 18–21.

Clift, R. (2006) 'Risk Management and Regulation in an Emerging Technology', in G. Hunt and M.D. Mehta (eds) *Nanotechnology: Risk, Ethics and Law*. London: Earthscan.

Cobb, M. (2005) 'Framing Effects on Public Opinion about Nanotechnology', *Science Communication*, 27 (2), 221–239.

Cobb, M.D. and J. Macoubrie (2004) 'Public Perceptions about Nanotechnology: Risks, Benefits and Trust', *Journal of Nanoparticle Research*, 6 (4), 395–405.

Collins, H.M. and T. Pinch (1998) *The Golem: What Everyone Should Know about Science*. Cambridge: Cambridge University Press.

Colvin, V.L. (2003) 'The Potential Environmental Impact of Engineered Nanoparticles', *Nature Biotechnology*, 21 (10), 1166–1170.

Condit, C.M. (2004) 'Science Reporting to the Public', *Canadian Medical Association Journal*, 170 (9), 1415–1416.

Cook, G., Pieri, E., and P.T. Robbins (2004) 'The Scientists Think and the Public Feels: Experts Perceptions of the Discourse of GM Food', *Discourse and Society*, 15 (4), 433–449.

Cooper, G. and M. Ebeling (2007) 'Epistemology, Structure and Urgency: The Sociology of Financial and Scientific Journalists', *Sociological Research Online*, 12 (3), http://www.socresonline.org.uk/12/3/8.html accessed 20 May 2008.

Cordis (2004) EU Policies for Nanosciences and Nanotechnologies. ftp://ftp.cordis.europa.eu/pub/nanotechnology/docs/eu_nano_policy_2004-07.pdf accessed 16 May 2008.

Corrigan, O. and A. Petersen (2008) 'UK Biobank: Bioethics as a Technology of Governance', in H. Gottweis and A. Petersen (eds) *Biobanks: Governance in Comparative Perspective*. London and New York: Routledge:

Cottle, S. (1998) 'Ulrich Beck, "Risk Society: and the Media: A Catastrophic View?', *European Journal of Communication*, 13 (1), 5–32.

Cottle, S. (ed.) (2003) *News, Public Relations and Power*. London: Sage.

Couldry, N, Livingstone, S., and T. Markham (2007) *Media Consumption and Public Engagement: Beyond the Presumption of Attention*. Basingstoke: Palgrave. http://www.lse.ac.uk/collections/media@lse/pdf/Public%20Connection%20 Report.pdf.accessed 4 June 2008.

Court E., Daar, A.S., Martin E., Acharya T., and P.A. Singer (2004) 'Will Prince Charles et al Diminish the Opportunities of Developing Countries in Nanotechnology?' http://www.nanotechweb.org/articles/society/3/1/1/1 accessed 16 May 2008.

Cranor, C. (ed.) (1994) *Are Genes Us? The Social Consequences of the New Genetics*. New Jersey: Rutgers University Press.

Crichton, M. (2002) *Prey*. New York: HarperCollins.

CRN (2008) 'What is Nanotechnology?', Centre for Responsible Nanotechnology http://www.crnano.org/whatis.htm accessed 6th June 2008.

CST (2007) *Nanosciences and Nanotechnologies: A Review of the Government's Progress on its Policy Commitments*. London: Council for Science and Technology.

Currall, S.C., King, E.B., Lane, N., Madera, J., and S. Turner (2006) 'What Drives Public Acceptance of Nanotechnology?', *Nature Nanotechnology*, 1 (3), 153–155.

Davis, A. (2000) 'Public Relations, News Production and Changing Patterns of Source Access in the British National Media', *Media, Culture and Society*, 22 (1), 39–59.

Dawkins, R. (1998) *Unweaving the Rainbow*. London: Penguin.

Dean, C. (2002) 'New Complications in Reporting on Science', *Nieman Reports*, 56 (3), 25–26.

Department of Innovation, Industry, Science and Research (2008) *National Nanotechnology Strategy*. January, Canberra: DIISR. http://www.innovation.gov.au/Section/Innovation/Documents/ NNSFeb08.pdf accessed 25 March 2008.

DTI/OST (2002) *New Dimensions for Manufacturing*: *A UK Strategy for Nanotechnology*. London: Department of Trade and. Industry/Office of Science and Technology.

Drexler, E. (2004) 'Nanotechnology: From Feynman to Funding', *Bulletin of Science Technology Society*, 24 (1), 21–27.

Drexler, E. (2008) 'Nanotechnology Overview' http://www.edrexler.com/p/04/03/0325nanoMeanings.html

Dunwoody, S. (1999) 'Scientists, Journalists and the Meaning of Uncertainty', in S.M. Friedman, S. Dunwoody, and C.L. Rogers (eds) *Communicating Uncertainty*: *Media Coverage of New and Controversial Science*. New Jersey: Lawrence Erlbaum Associates.

Ebeling, M. (2008) 'Mediating Uncertainty: Communicating the Financial Risks of Nanotechnologies', *Science Communication*, 29 (3), 335–361.

Einsiedel, E. and L. Goldenberg (2004) 'Dwarfing the Social? Nanotechnology Lessons from the Biotechnology Front', *Bulletin of Science, Technology and Society*, 24 (1), 28–33.

Eisend, M. (2002) 'The Internet as a New Medium for the Sciences?', *Online Information Review*, 26 (5), 307–317.

Eldridge, J. and J. Reilly (2003) 'Risk and Relativitiy: BSE and the British Media', in N. Pidgeon, R.E. Kasperson, and P. Slovic (eds) *The Social Amplification of Risk*. Cambridge: Cambridge University Press.

ETC Group (2003) *The Big Down*: *Atomtech*: *Technologies Converging at the Nanoscale*, Winnipeg, Canada: ETC Group. http://www.etcgroup.org/upload/publication/171/01/thebigdown.pdf accessed 16 May 2008.

Europa (2008) 'European Commission adopts Code of Conduct for Responsible Nanosciences and Nanotechnologies Research', Press Release. http://europa.eu/rapid/pressReleasesAction.do?reference=IP/08/193&format=HTML&aged=0&language=EN&guiLanguage=en accessed 25 March 2008.

European Commission (2006) *Europeans and Biotechnology in 2005*: *Patterns and Trends*. Brussels: Directorate General Press and Communication.

Evans, G. and J. Durant, (1995) 'The Relationship between Knowledge and Attitudes in the Public Understanding of Science in Britain', *Public Understanding of Science*, 4 (1), 57–74.

Evans, R. and A. Plows (2007) 'Listening without Prejudice?: Re-discovering the Value of the Disinterested Citizen', *Social Studies of Science*, 37 (6), 827–853.

Evans, W. and S.H. Priest (1995) 'Science Content and Social Context', *Public Understanding of Science*, 4 (4), 327–340.

Eveland, W.P. and S. Dunwoody (1998) 'Users and Navigation Patterns of a Science World Wide Web Site for the Public', *Public Understanding of Science*, 7 (4), 285–311.

Faber, B. (2005) *Popularizing Nanoscience*: *The Public Rhetoric of Nanotechnology, 1986–1999*. http://www.people.clarkson.edu/~faber/pubs/nano.tech.tcq.3.0.doc accessed 28 February 2008.

Faber, B. (2006) 'Popularizing Nanoscience: The Public Rhetoric of Nanotechnology, 1986–1999', *Technical Communication Quarterly*, 15 (2), 141–169.

Flatow, I. (1997) 'Magazine Style', in D. Blum and M. Knudson (eds) *A Field Guide for Science Writers*. New York: Oxford University Press.

Flynn, J., Slovic, P., and H. Kunreuther (eds) (2001) *Risk, Media and Stigma: Understanding Public Challenges to Modern Science and Technology*. London: Earthscan.

Friedhoff, S. (2002) 'Rethinking the Science Beat', *Nieman Reports*, 56 (3), 23–25.

Friedman, S.M. (1986) 'The Journalist's World', in S.M. Friedman, S. Dunwoody, and C.L. Rogers (eds) *Scientists and Journalists*. New York: Free Press.

Friedman, S.M. and B.P. Egolf (2005) 'Nanotechnology, Risks and the Media', *IEEE Technology and Society Magazine*, 24 (4), 5–11.

Friedman, S.M., Dunwoody, S., and C.L. Rogers (eds) (1986) *Scientists and Journalists*. Washington: AAAS.

Friedman, S., Dunwoody, S., and C.L. Rogers (eds) (1999) *Communicating Uncertainty: Media Coverage of New and Controversial Science*. Mahwah, NJ: Lawrence Erlbaum.

Friends of the Earth Australia (2008a) *Out of the Laboratory and on to Our Plates: Nanotechnology in Food and Agriculture*. Friends of the Earth Australia, Europe and USA. http://nano.foe.org.au/filestore2/download/228/Nanotechnology%20in%20food%20and%20agriculture%20-%20text%20only%20version.pdf accessed 22 April 2008.

Friends of the Earth Australia (2008b) *Mounting Evidence that Carbon Nanotubes may be the New Asbestos*. Friends of the Earth Australia. http://nano.foe.org.au/filestore2/download/265/Mounting%20evidence%20that%20carbon%20nanotubes%20may%20be%20the%20new%20asbestos%20%20August%202008.pdf accessed 19 September 2008.

Gaskell, G. (2004) 'Science Policy and Society: The British Debate over GM Agriculture', *Current Opinion in Biotechnology*, 15 (3), 241–245.

Gaskell, G., Ten Eyck, T., Jackson, J., and G. Veltri (2005) 'Imagining Nanotechnology: Cultural Support for Technological Innovation in Europe and the United States', *Public Understanding of Science*, 14 (1), 81–90.

Gavelin, K., Wilson, R., and R. Doubleday (2007) *Democratic Technologies? The Final Report of the Nanotechnologies Engagement Group (NEG)*. London: Involve.

Gibson, I. (2003) 'Make Truth the Target', *Times Higher Education Supplement*, 2 May.

Gieryn, T.F. (1999) *Cultural Boundaries of Science: Credibility on the Line*. Chicago: University of Chicago Press.

Goffman, E. (1974) *Frame Analysis*. New York: Harper and Row.

Goodell, R. (1986) 'How to Kill a Controversy: The Case of Recombinant DNA', in S. Friedman, S. Dunwoody, and C. Rogers (eds) *Scientists and Journalists: Reporting Science as News*. New York: Free Press.

Gorss, J. and B. Lewenstein (2005) 'The Salience of Small: Nanotechnology Coverage in the American Press, 1986–2004', Paper presented at International Communication Association Conference, 27 May.

Gottweis, H. and A. Petersen (2008) 'Biobanks and Governance: An Introduction', in H. Gottweis and A. Petersen (eds) *Biobanks: Governance in Comparative Perspective*. London and New York: Routledge.

Greenberg, J. (1997) 'Using Sources', in D. Blum and M. Knudson (eds) *A Field Guide for Science Writers*. New York: Oxford University Press.

Greenberg, M.R., Sachsman, D.B., Sandman, P.M., and K.L. Salome (1989) 'Network Evening News Coverage of Environmental Risk', *Risk Analysis*, 9 (1), 119–126.

Gregory, J. and S. Miller (1998) *Science in Public*. Cambridge: Perseus.

Gregory, J. and S. Miller (2001) 'Caught in the Crossfire: The Public's Role in the Science Wars', in J.A. Labinger and H. Collins (eds) *The One Culture?* Chicago: University of Chicago Press.

Gregory, R., Flynn, J., and P. Slovic (2001) 'Technological Stigma', in J. Flynn, P. Slovic and H. Kunreuther (eds) *Risk, Media and Stigma: Understanding Public Challenges to Modern Science and Technology*. London: Earthscan.

Gunter, B., Kinderlerer, J., and D. Beyleveld (1999) 'The Media and Public Understanding of Biotechnology: A Survey of Scientists and Journalists', *Science Communication*, 20 (4), 373–394.

Gurabardhi, Z., Gutteling, J.M., and M. Kuttschreuter (2005) 'An Empirical Analysis of Communication Flow, Strategy and Stakeholders' Participation in the Risk Communication Literature 1988–2000', *Journal of Risk Research*, 8 (6), 499–511.

Hall, S. (1981) 'The Determinations of News Photographs', in S. Cohen and J. Young (eds) *The Manufacture of News* (Revised Edition). London: Constable.

Handy, R.D. and B.J. Shaw (2007) 'Toxic Effects of Nanoparticles and Nanomaterials: Implications for Public Health, Risk Assessment and the Public Perception of Nanotechnology', *Health, Risk and Society*, 9 (2), 125–144.

Hannah, W. and P.B. Thompson (2008) 'Nanotechnology, Risk and the Environment: A Review', *Journal of Environmental Monitoring*, 10 (3), 291–300.

Hansen, A. (ed.) (1993) *The Mass Media and Environmental Issues*. Leicester: Leicester University Press.

Hansen, A. (1994) 'Journalistic Practices and Science Reporting in the British Press', *Public Understanding of Science*, 3 (2), 111–134.

Hansen, A. (2000) 'Claims-making and Framing in British Newspaper Coverage of the "Brent Spar" Controversy', in S. Allan, B. Adam, and C. Carter (eds) *Environmental Risks and the Media*. London: Routledge.

Hansen, A., Cottle, S., Negrine, R., and C. Newbold (1998) *Mass Communication Research Methods*. London: Macmillan.

Hansson, S.O. (2004) 'Great Uncertainty about Small Things', *Techne*, 8 (2), 26–35.

Hargreaves, I. and G. Ferguson (2000) *Who's Misunderstanding Whom? Bridging the Gulf of Understanding between the Public, the Media and Science*. Swindon: Economic and Social Research Council.

Hargreaves, I., Lewis, J., and T. Speers (2003) *Towards a Better Map: Science, the Public and the Media*. Swindon: Economic and Social Research Council.

Harris, R.F. (1997) 'Toxics and Risk Reporting', in D. Blum and M. Knudson (eds) *A Field Guide for Science Writers*. New York: Oxford University Press.

Heckl, W. (2007) *Nanodialogue: Recommendations to Achieve Sustainable Governance and Social Acceptance*. EuroNanoForum 2007, Nanotechnology in Industrial Applications. 19–21 June 2007, Düsseldorf, Germany: CCD.

Hermida, A. (2006) 'Young Challenge Mainstream Media', *BBC News Online*, 3 May 2006. http://news.bbc.co.uk/1/hi/technology/4962794.stm accessed 4 June 2006.

Hermida, A. (2007) 'Reimagining Science Journalism', Future Directions in Science Journalism conference, University of British Columbia, 9–10 November.

Hett, A. (2004) *Nanotechnology: Small Matter, Many Unknowns*. Zurich: Swiss Re Insurance Company.

Hilgartner, S. (1990) 'The Dominant View of Popularization: Conceptual Problems, Political Uses', *Social Studies of Science*, 20 (3), 519–539.

Hilgartner, S. (2000) *Science on Stage: Expert Advice as Public Drama*. Stanford: Stanford University Press.

Hilgartner, S. and C.L. Bosk (1988) 'The Rise and Fall of Social Problems: A Public Arenas Model', *American Journal of Sociology*, 94 (1), 53–78.

HM Government (2005a) *Response to the Royal Society and Royal Academy of Engineering Report: Nanoscience and Nanotechnologies: Opportunities and Uncertainties*. February 2005. London: Department of Trade and Industry.

HM Government (2005b) *The Government's Outline Programme for Public Engagement on Nanotechnologies*. London: HM Government in Consultation with the Devolved Administrations. http://www.ost.gov.uk/policy/issues/programme12.pdf accessed 16 May 2008.

HMSO (2000) *Science and Society; Third Report of the Session 1999–2000*. London: HM Stationery Office.

HMT, DfES, and DTI (2004) *Science and Innovation Investment Framework 2004–2014*. London: HM Stationery Office.

Hoet, P.H.M., Bruske-Hohlfield, I., and O.V. Salata (2004) 'Nanoparticles – Known and Unknown Health Risks', *Journal of Nanobiotechnology*, 2 (12). http://www.jnanobiotechnology.com/content/2/1/12

Hornig, S. (1993) 'Reading Risk: Public Response to Print Media Accounts of Technological Risk', *Public Understanding of Science*, 2 (2), 95–109.

Hornig Priest, S. (2001) 'Cloning: A Study in News Production', *Public Understanding of Science*, 10 (1), 59–69.

Hornig Priest, S. (2005) 'Risk Reporting: Why can't they ever get it Right?', in S. Allan (ed.) *Journalism: Critical Issues*. Buckingham: Open University Press.

Hotz, R.L. (2002) 'The Difficulty of Finding Impartial Sources in Science', *Nieman Reports*, 56 (3), 6–7.

House of Commons Science and Technology Committee (2004) *Too Little, Too Late? Government Investment in Nanotechnology: Fifth Report of Session 2003–4. Volume One*. London: The Stationery Office.

Human Genome Project Information (2008) *Ethical, Legal and Social Issues*. http://www.ornl.gov/sci/techresources/Human_Genome/elsi/elsi.shtml accessed 28 February 2008.

Hunt, G. (2006) 'Nanotechnologies and Society in Europe', in G. Hunt and M.D. Mehta (eds) *Nanotechnology: Risk, Ethics and Law*. London: Earthscan.

Hunt, G. and M.D. Mehta (eds) (2006) *Nanotechnology: Risk, Ethics and Law*. London: Earthscan.

Iredale, R. and M. Longley (2000) 'Reflections on Citizens' Juries: The Case of the Citizens' Jury on Genetic Testing for Common Disorders', *Journal of Consumer Studies and Home Economics*, 24 (1), 41–47.

Irwin, A. (1995) *Citizen Science: A Study of People, Expertise and Sustainable Development*. London: Routledge.

Irwin, A. (2001) 'Constructing the Scientific Citizen: Science and Democracy in Biosciences', *Public Understanding of Science*, 10 (1), 1–18.

Irwin, A. and Michael, M. (2003) *Science, Social Theory and Public Knowledge*. Maidenhead: Open University Press.

Jarmul, D. (1997) 'Op-ed Writing', in D. Blum and M. Knudson (eds) *A Field Guide for Science Writers*. New York: Oxford University Press.

Jasanoff, S. (2005) *Designs on Nature: Science and Democracy in Europe and the United States*. Princeton, NJ and Oxford: Princeton University Press.

Jensen, E. (2008) 'The Dao of Human Cloning: Utopian/Dystopian Hype in the British Press and Popular Films', *Public Understanding of Science*, 17 (2), 123–143.

Joly, P.B. and A. Kaufmann, (2008) 'Lost in Translation: The Need for "Upstream Engagement" with Nanotechnology on Trial', *Science as Culture*, 17 (3), 225–247.

Jones, R. (2008) 'The Economy of Promises', *Nature Nanotechnology*, 3, 65–66. http://www.nature.com/nnano/index.html

Kasperson, R., Jhaveri, N., and J.X. Kasperson (2001) 'Stigma and the Social Amplification of Risk: Toward a Framework of Analysis', in J. Flynn, P. Slovic, and H. Kunreuther (eds) *Risk, Media and Stigma: Understanding Public Challenges to Modern Science and Technology*. London: Earthscan.

Kasperson, R.E., Renn, O., Slovic, P., Brown, H.S., Emel, J., Goble, R., Kasperson, J.X. and S.J. Ratick (1988) 'The Social Amplification of Risk: A Conceptual Framework', *Risk Analysis*, 8 (2), 178–187.

Kiernan, V. (2006) *Embargoed Science*. Urbana: University of Illinois Press.

Kitzinger, J. (1999) 'Researching Risk and the Media', *Health, Risk and Society*, 1 (1), 55–69.

Kitzinger, J. (2006) 'The Role of the Media in Public Engagement with Science', in J. Turney (ed.) *Engaging Science: Thoughts, Deeds, Analysis and Action*. London: Wellcome Trust.

Kitzinger, J. and J. Reilly (1997) 'The Rise and Fall of Risk Reporting: Media Coverage of Human Genetics Research, "False Memory Syndrome" and "Mad Cow Disease" ', *European Journal of Communication*, 12 (3), 319–350.

Kitzinger, J., Henderson, L., Smart, A., and J. Eldridge (2003) *Media Coverage of the Social and Ethical Implications of Human Genetic Research*. Final Report to the Wellcome Trust, February. Award no: GR058105MA.

Koolstra, C.M., Bos, M.J.W., and I.E. Vermeulen (2006) 'Through which Medium should Science Information Professionals Communicate with the

Public: Television or the Internet?', *Journal of Science Communication*, 5 (3), 1–8.

Kulinowski, K. (2004) 'Nanotechnology: from "Wow" to "Yuck"?', *Bulletin of Science, Technology and Society*, 24 (1), 13–20.

Kulinowski, K. (2006) 'Nanotechnology: From "Wow" to "Yuck"?', in G. Hunt and M.D. Mehta (eds) *Nanotechnology: Risk, Ethics and Law*. London: Earthscan.

LaMonica, M. (2008) 'IBM, Saudis Partner on "Green" Nanotech Lab', Green Tech Blog http://www.news.com/8301-11128_3-9884156 54.html?part=rss&tag=feed&subj=GreenTechblog accessed 25 March 2008.

Latour, B. (1987) *Science in Action*. Harvard, MA: Cambridge University Press.

Lee, C. and D.A. Scheufele (2006) 'The Influence of Knowledge and Deference toward Scientific Authority: A Media Effects Model for Public Attitudes toward Nanotechnology', *Journalism and Mass Communication Quarterly*, 83 (4), 819–834.

Lehoux, P. (2006) *The Problem of Health Technology: Policy Implications for Modern Health Care Systems*. New York: Routledge

Lewenstein, B.V. (1995) 'Science and Media', in S. Jasanoff, G.E. Markle, J.C. Petersen and T. Pinch (eds) *Handbook of Science and Technology Studies*. Thousand Oaks: Sage.

Lewenstein, B.V., Radin, J., and J. Diels (2004) 'Nanotechnology in the Media: A Preliminary Analysis', in W.S. Bainbridge (ed.) *Societal Implications of Nanoscience and Nanotechnology II: Maximizing Human Benefit* (Report of the National Nanotechnology Initiative Workshop, 3–5 December 2003, Arlington, VA), Washington DC: National Science & Technology Council and National Science Foundation.

Lichtenberg, J. and D. MacClean (1991) 'The Role of the Media in Risk Communication', in R. Kasperson and P. Stallen (eds) *Communicating Risks to the Public: International Perspectives*. London: Kluwer Academic.

Logan, R.A. (1991) 'Popularization versus Secularization: Media Coverage of Health', in L. Wilkins and P. Patterson (eds) *Risky Business: Communicating Issues of Science, Risk and Public Policy*. New York: Greenwood.

Losch, A. (2006) 'Anticipating the Futures of Nanotechnology: Visionary Images as Means of Communication', *Technology Analysis and Strategic Management*, 18 (3), 393–409.

Lupton, D. (2004) 'A Grim Health Future: Food Risks in the Sydney Press', *Health, Risk and Society*, 6 (2), 187–200.

Lynch, J. and C.M. Condit (2006) 'Genes and Race in the News', *American Journal of Health Behaviour*, 30 (2), 125–135.

Maason, S and P. Weingart (2000) *Metaphors and the Dynamics of Knowledge*. New York: Routledge.

McCallum, D.B., David, B., Hammond, S.L., and V.T. Covello (1991) 'Communicating about Environmental Risks: How the Public uses and Perceives Information Sources', *Health Education Quarterly*, 18 (3), 349–361.

Macnaghten, P. and J. Urry (1998) *Contested Natures*. London: Sage.

Macoubrie, J. (2005) *Informed Public Perceptions of Nanotechnology and Trust in Government.* Project on Emerging Technologies at the Woodrow Wilson International Center for Scholars. http://www.wilsoncenter.org/index.cfm?fuseaction=events.event_summary &event_i=143410 accessed 2 August 2007.

Malone, R.E., Boyd, E., and L.A. Bero (2000) 'Science in the News: Journalists' Constructions of Passive Smoking as a Social Problem', *Social Studies of Science*, 30 (5), 713–735.

Manning, P. (2001) *News and News Sources: A Critical Introduction.* London: Sage.

Market Attitude Research Services (2007) *Final Report: Australian Community Attitudes Held about Nanotechnology – Trends 2005 to 2007.* Sydney: Market Attitude Research Services Pty Ltd.

Marks, L.A., Kalaitzandonakes, N., Wilkins, L., and L. Zakharova (2007) 'Mass Media Framing of Biotechnology News', *Public Understanding of Science*, 16 (2), 183–203.

Masami, M., Hunt, G., and O. Masayuki (2006) 'Nanotechnologies and Society in Japan', in G. Hunt, G. and M.D. Mehta (eds) *Nanotechnology: Risk, Ethics and Law.* London: Earthscan.

Massoli, L. (2007) 'Science on the Net: An Analysis of the Websites of the European Public Research Institutions', *Journal of Science Communication*, 6 (3), 1–16.

Mayer, S. (2002) 'From Genetic Modification to Nanotechnology: The Dangers of "Sound Science" ', in T. Gilland (ed.) *Science: Can We Trust the Experts?* London: Hodder and Stoughton.

Mehta, M.D. (2004) 'From Biotechnology to Nanotechnology: What can we Learn from Earlier Technologies?', *Bulletin of Science, Technology and Society*, 24 (1), 34–39.

Mehta, M.D. (2005) *Risky Business: Nuclear Power and Public Protest in Canada.* Lexington, MD: Lanham.

Meili, C. (2005) 'The "Ten Commandments" of Nano Communication – Or how to Deal with Public Perception.' Innovation Society. http://www.innovationsgesellschaft.ch/images/publikationen/Ten%20Commandments%20of%20Nano-Communication.pdf accessed 16 May 2008.

Michael, M. (2002). 'Comprehension, Apprehension, and Prehension: Heterogeneity and the Public Understanding of Science', *Science, Technology and Human Values*, 27(3), 357–370.

Michelson, E.S. and D. Rejeski (2006) *Falling Through the Cracks? Public Perception, Risk, and the Oversight of Emerging Nanotechnologies.* IEEE, Technology and Society International Symposium.

Miksanek, T. (2001) 'Microscopic Doctors and Molecular Black Bag: Science Fiction's Prescription for Nanotechnology and Medicine', *Literature and Medicine*, 21 (1), 55–70.

Miller, D. and J. Reilly (1994) *Food Scares in the Media.* Glasgow University Media Group. http://homepages.strath.ac.uk/~ his04105/publications/Foodscares.html accessed 16 May 2008.

Miller, D. and B. Dinan (2000) 'The Rise of the PR Industry in Britain: 1979–98', *European Journal of Communication*, 15 (1), 5–35.

Miller, M. and B.P. Riechert (2000) 'Interest Group Strategies and Journalistic Norms: News Media Framing of Environmental Issues', in S. Allan, B. Adam, and C. Carter (eds) *Environmental Risks and the Media*. London: Routledge.

Miller, S. (2001) 'Public Understanding of Science at the Crossroads', *Public Understanding of Science*, 10 (1), 115–120.

MORI (2000) *The Role of Scientists in Public Debate*. London: Wellcome Trust.

Murdock, G. (2004) 'Popular Representations and Postnormal Science: The Struggle over Genetically Modified Foods', in S. Braman (ed.) *Biotechnology and Communication: The Meta-Technologies of Information*. Mahwah, NJ: Lawrence Erlbaum.

Murdock, G., Petts J., and T. Horlick-Jones (2003) 'After Amplification: Rethinking the Role of the Media in Risk Communication', in N. Pidgeon, R.E. Kasperson, and P. Slovic (eds) *The Social Amplification of Risk*. Cambridge: Cambridge University Press.

Mythen, G. (2004) *Ulrich Beck: A Critical Introduction to the Risk Society*. London: Pluto Press.

Mythen, G. (2007) 'Reappraising the Risk Society Thesis: Telescopic Sight or Myopic Vision?', *Current Sociology*, 55 (6), 793–813.

Nano2Life (2008) Nano2Life website http://www.nano2life.de/content.php?id=2 accessed 25 March 2008.

Nanoforum Report (2004) *The 4th Nanoforum Report: Benefits, Risks, Ethical, Legal and Social Aspects of Nanotechnology*, Nanoforum.org: European Nanotechnology Gateway, June. http://www.nanoforum.org/ accessed 16 May 2008.

Nanoforum (2008) http://www.nanoforum.org/ accessed 7 June 2008.

Nanologue (2006) *Nanologue: Opinions of Ethical, Legal and Social Aspects of Nanotechnologies*. Wuppertal: Wuppertal Institute, Forum for the Future and Triple Innova, 1–117.

Nelkin, D. (1987) *Selling Science: How the Press Covers Science and Technology*. New York: W. H Freeman and Company.

Nelkin, D. (1995) *Selling Science: How the Press covers Science and Technology*. New York: WH Freeman.

Newspaper Marketing Agency (2005) 'Newspaper Crib Sheet'. http://www.nmauk.co.uk/nma/do/live/factsAndFigures accessed 20 May 2008.

Nisbet, M.C., Brossard, D., and A. Kroepsch (2003) 'Framing Science: The Stem Cell Controversy in the Age of Press/Politics', *Harvard International Journal of Press/Politics*, 8 (2), 36–70.

Nisbet, M.C. and B.V. Lewenstein (2002) 'Biotechnology and the American Media: The Policy Process and the Elite Press, 1970 to 1999', *Science Communication*, 23 (4), 359–359.

Nisbet, M.C. and C. Mooney (2007) 'Framing Science', *Science*, 316 (5821), 6 April, 56.

Nisbet, M.C. and M. Huge (2006) 'Attention Cycles and Frames in the Plant Biotechnology Debate: Managing Power and Participation through the

Press/Policy Connection', *Harvard International Journal of Press/Politics*, 11 (2), 3–40.

Nisbet, M.C. and M. Huge (2007) 'Where do Science Debates come from? Understanding Attention Cycles and Framing?', in D. Brossard, J. Shanahan, and T.C. Nesbitt (eds)*The Public, the Media and Agricultural Biotechnology*. London: CABI.

Nowotny, H., Scott, P., and M. Gibbons (2001) *Re-Thinking Science, Knowledge and the Public in an Age of Uncertainty*. Cambridge: Polity Press.

NSF (2006) *Science and Engineering Indicators 2006: Volume One*. Arlington: National Science Foundation.

NSF (2008) National Science Foundation http://www.nsf.gov/crssprgm/nano/reports/omb_nifty50.jsp accessed 6 June 2008.

OST and The Wellcome Trust (2000) *Science and the Public: A Review of Science Communication and Public Attitudes to Science in Britain*. London: Department of Trade and Industry and the Wellcome Trust.

Oud, M. and I. Malsch (2003) *Socio-Economic Report on Nanotechnology and Smart Materials for Medical Devices*, Nanoforum. http://www.nanoforum.org/ accessed 14 May 2008.

Park, B. (2007) 'Current and Future Applications of Nanotechnology', in R.E. Hester and R.M. Harrison (eds) *Nanotechnology: Consequences for Human Health and the Environment*. Cambridge: Royal Society of Chemistry.

Paul, D. (2004) 'Spreading Chaos: The Role of Popularizations in the Diffusion of Scientific Ideas', *Written Communication*, 21 (1), 32–68.

Pearson, G., Pringle, S.M., and J.N. Thomas (1997) 'Scientists and the Public Understanding of Science', *Public Understanding of Science*, 6 (3), 279–289.

Pellechia, M.G. (1997) 'Trends in Science Coverage: A Content Analysis of Three US Newspapers', *Public Understanding of Science*, 6 (1), 49–68.

Pense, C.M. and S.H. Cutcliffe (2007) 'Risky Talk: Framing the Analysis of the Social Implications of Nanotechnology', *Bulletin of Science, Technology and Society*, 27 (5), 349–366.

Perlman, D. (1997) 'Introduction', in D. Blum and M. Knudson (eds) *A Field Guide for Science Writers*. New York: Oxford University Press.

Peters, H.P. (1999) 'The Interaction of Journalists and Scientific Experts: Co-operation and Conflict between Two Professional Cultures', in E. Scanlon, E. Whitelegg, and S. Yates (eds) *Communicating Science Contexts and Channels*. London: Routledge.

Petersen, A. (2001) 'Biofantasies: Genetics and Medicine in the Print News Media', *Social Science and Medicine*, 52 (8), 1255–1268.

Petersen, A. (2002) 'Replicating our Bodies, Losing Ourselves: News Media Portrayals of Human Cloning in the Wake of Dolly', *The Body & Society*, 8 (4), 71–90.

Petersen, A. (2005) 'The Metaphors of Risk: Biotechnology in the News', *Health, Risk and Society*, 7 (3), 203–208.

Petersen, A., Allan, S., Anderson, A., and C. Wilkinson (2009) 'Opening the Black Box: Scientists' Views on the Role of the Mass Media in the Nanotechnology Debate', *Public Understanding of Science*.

Petersen, A., Anderson, A., and S. Allan (2005) 'Science Fiction/Science Fact: Medical Genetics in Fictional and News Stories', *New Genetics and Society*, 24 (3), 337–353.

Petersen, A. and A. Anderson (2008) 'A Question of Balance or Blind Faith?: Scientists' and Science Policymakers' Representations of the Benefits and Risks of Nanotechnologies', *NanoEthics*, 1 (3), 243–256.

Petersen, A., Anderson, A., Wilkinson, C., and S. Allan (2007) 'Editorial: Nanotechnologies, Risk and Society', *Health, Risk and Society*, 9 (2), 117–124.

Petit, C. (1997) 'Covering Earth Sciences', in D. Blum and M. Knudson (eds) *A Field Guide for Science Writers*. New York: Oxford University Press.

Petts, J., Horlick-Jones, T., and G. Murdock (2001) *Social Amplification of Risk: The Media and the Public, Contract Research Report*. London: Health and Safety Executive.

Petts J. and S. Niemeyer (2004) 'Health Risk Communication and Amplification: Learning from the MMR Vaccination Controversy', *Health, Risk and Society*, 6 (1), 7–23.

Pew Internet Project/Exploratorium (2006) *The Internet as a Resource for News and Information about Science*, http://www.pewinternet.org/ accessed 16 May 2008.

Philo, G. (ed.) (1999) *Message Received*. Harlow: Longman.

Phoenix, N. and E. Drexler (2004) 'Safe Exponential Manufacturing', *Nanotechnology*, 15 (8), 869–872.

Pidgeon, N., Kasperson, R.E., and P. Slovic (eds) (2003) *The Social Amplification of Risk*. Cambridge: Cambridge University Press

Pinholster, G. and C.O'Malley (2006) 'EurekAlert! Survey Confirms Challenges for Science Communicators in the Post-Print Era', *Journal of Science Communication*, 5 (3), 1–12.

POST (2001) *Open Channels: Public Dialogue in Science and Technology*. London: Parliamentary Office of Science and Technology.

Powell, M.C. (2007) 'New Risk or Old Risk, High Risk or No Risk?: How Scientists' Standpoints Shape their Nanotechnology Risk Frames', *Health, Risk and Society*, 9 (2), 173–190.

Priest, S. (2006) 'The North American Opinion Climate for Nanotechnology and its Products: Opportunities and Challenges', *Journal of Nanoparticle Research*, 8 (4), 563–568.

Randles, S., Dewick, P., Loveridge, D., and J.C. Schmidt (2008) 'Nano-worlds as Schumpeterian Emergence and Polanyian Double-Movements', *Analysis & Strategic Management*, 20 (1), 1–11.

RCUK/DIUS (2008) *Public Attitudes to Science 2008: A Survey*. London: People, Science and Policy Limited. http://www.rcuk.ac.uk/cmsweb/downloads/rcuk/scisoc/pas08.pdf/ accessed 16 May 2008.

Reed, R. (2001) '(Un) Professional Discourse? Journalists and Scientists' Stories about Science in the Media', *Journalism: Theory, Practice and Criticism*, 2 (3), 279–298.

Reed, R. and G.F. Walker (2002) 'Listening to Scientists and Journalists', *Nieman Reports*, 56 (3), 45–46.

Renn, O., Burns, W.J., Kasperson, J.X., Kasperson, R.E., and P. Slovic (1992) 'The Social Amplification of Risk: Theoretical Foundations and Empirical Applications', *Risk Analysis*, 48 (4), 137–160.

Renn, O. and M.C. Roco (2006) 'Nanotechnology and the Need for Risk Governance', *Journal of Nanoparticle Research*, 8 (2), 153–191.

Rensberger, B. (1997) 'Covering Science for Newspapers', in D. Blum and M. Knudson (eds) *A Field Guide for Science Writers*. New York: Oxford University Press.

Richardson, K. (2001) 'Risk News in the World of Internet Newsgroups', *Journal of Sociolinguistics*, 5 (1), 50–72.

Roco, M. and W.S. Bainbridge (2001) *Societal Implications of Nanoscience and Nanotechnology*. National Science Foundation. http://www.wtec.org/loyola/nano/NSET.Societal.Implications/ accessed 16 May 2008.

Rogers-Hayden, T. and N. Pidgeon (2007) 'Moving Engagement "Upstream"? Nanotechnologies and the Royal Society and Royal Academy of Engineering's Inquiry', *Public Understanding of Science*, 16 (3), 345–364.

Ropeik, D. (1997) 'Reporting News', in D. Blum and M. Knudson (eds) *A Field Guide for Science Writers*. New York: Oxford University Press.

Rose, H. (2000) 'Risk, Trust and Scepticism in the Age of New Genetics', in B. Adam, U. Beck, and J. Van Loon (eds) *The Risk Society and Beyond*: *Critical Issues for Social Theory*. London: Sage.

Rowe, G. and L.J. Frewer (2000) 'Public Participation Methods: A Framework for Evaluation', *Science, Technology and Human Values*, 25 (1), 3–29.

Rowell, A. (2003) *Don't Worry it's Safe to Eat*: *The True Story of GM Food, BSE and Foot and Mouth*. London: Earthscan.

Royal Society (1985) *The Public Understanding of Science*. London: Royal Society.

Royal Society (2006) *Science Communication*: *Survey of Factors Affecting Science Communication by Scientists and Engineers*. London: Royal Society.

Royal Society (2007) 'Communication Skills and Media Training Courses', at http://www.royalsoc.ac.uk/page.asp?id=1151 Accessed 5 September 2007.

Royal Society and Royal Academy of Engineering (RS/RAE) (2004) *Nanoscience and Nanotechnologies*: *Opportunities and Uncertainties Report*. London: The Royal Society. http://www.nanotec.org.uk/finalReport.htm accessed 16 May 2008.

Royal Society and Royal Academy of Engineering (2007) *Joint Academies Initial Response to the Council for Science and Technology's Review of Government's Progress on Nanotechnologies*. London: Royal Society.

Ryan, C. (1991) *Prime Time Activism*: *Media Strategies for Grassroots Organising*. Boston, MA: South End Press.

Salisbury, D.F. (1997) 'Colleges and Universities', in D. Blum and M. Knudson (eds) *A Field Guide for Science Writers*. New York: Oxford University Press.

Sandler, R. (2007) 'Nanotechnology and Social Context', *Bulletin of Science, Technology and Society*, 27 (6), 446–454.

Sandler, R. and W.D. Kay (2006) 'The GMO-Nanotech (Dis)Analogy?', *Bulletin of Science, Technology and Society*, 26 (1), 57–62.

Schanne, M. and W. Meier (1992) 'Media Coverage of Risk', in J. Durant (ed.) *Museums and the Public Understanding of Science*. London: Science Museum Publications.

Scheufele, D.A. and B.V. Lewenstein (2005) 'The Public and Nanotechnology: How Citizens make sense of Emerging Technologies', *Journal of Nanoparticle Research*, 7 (6), 659–677.

Scheufele, D.A., Corley, E.A., Dunwoody, S., Shih, T., Hillback, E., and D. Guston (2007) 'Scientists Worry about some Risks more than the Public', *Nature Nanotechnology*, 2 (12), 732–734.

Schummer, J. (2004) 'Societal and Ethical Implications of Nanotechnology: Meanings, Interest Groups and Social Dynamics', *Techne*, 8 (2), 56–87.

Schummer, J. (2005) 'Reading Nano: The Public Interest in Nanotechnology as Reflected in Purchase Patterns of Books', *Public Understanding of Science*, 14 (2), 163–183.

Science Media Centre (2002) *Consultation Report*. London: Science Media Centre.

Selin, C. (2007) 'Expectations and the Emergence of Nanotechnology', *Science, Technology & Human Values*, 32 (2), 196–220.

Singer, E. and P. Endreny (1987) 'Reporting Hazards: Their Benefits and their Costs', *Journal of Communication*, 37 (3), 10–26.

Smalley, R.E. (2008) Richard Smalley Institute for Nanoscale Science and Technology 'What is Nanotechnology' http://cnst.rice.edu/nano.cfm accessed 6th June 2008.

Snow, C.P. (1965) *The Two Cultures: A Second Look*. Cambridge: Cambridge University Press.

Stallings, R.A. (1990) 'Media Discourse and the Social Construction of Risk', *Social Problems*, 37 (1), 80–95.

Stephens, L.F. (2004) 'News Narratives about Nano: How Journalists and the News Media are Framing Nanoscience and Nanotechnology Initiatives and Issues'. Paper presented to Imaging and Imagining Nanoscience and Engineering Conference, University of South Carolina, 3–7 March.

Stephens, L.F. (2005) 'News Narratives about Nano S & T in Major U.S. and Non-U.S. Newspapers', *Science Communication*, 27 (2), 175–199.

Stocking, S.H. (1999) 'How Journalists Deal with Scientific Uncertainty', in S.M. Friedman, S. Dunwoody, and C.L. Rogers (eds) *Communicating Uncertainty: Media Coverage of New and Controversial Science*. New Jersey: Lawrence Erlbaum.

Te Kulve, H.T. (2006) 'Evolving Repertoires: Nanotechnology in Daily Newspapers in the Netherlands', *Science as Culture*, 15 (4), 367–382.

Theodore, L. and R.G. Kunz (2005) *Nanotechnology: Environmental Implications and Solutions*. New Jersey: John Wiley.

Thirlaway, K. and D. Heggs (2005) 'Interpreting Risk Messages: Women's Responses to a Health Story', *Health, Risk and Society*, 7 (2), 107–121.

Throne-Holst, H. and E.A. Stø (2008) 'Who should be Precautionary? Governance of Nanotechnology in the Risk Society', *Technology Analysis & Strategic Management*, 20 (1), 99–112.

Thurs, D.P. (2007) 'Tiny Tech, Transcendent Tech: Nanotechnology, Science Fiction, and the Limits of Modern Science Talk', *Science Communication*, 29 (1), 65–95.

Toner, M. (1997) 'Introduction', in D. Blum and M. Knudson (eds) *A Field Guide for Science Writers*. New York: Oxford University Press.

Tourney, C. (2004) 'Narratives for Nanotech: Anticipating Public Reactions to Nanotechnology', *Techne*, 8 (2), 88–116.

Toumey, C.P. (1996) 'Conjuring Science in the Case of Cold Fusion', *Public Understanding of Science*, 5 (2), 121–133.

Trafford, A. (1997) 'Critical Coverage of Public Health and Government', in D. Blum and M. Knudson (eds) *A Field Guide for Science Writers*. New York: Oxford University Press.

Treise, D., Walsh-Childers, K., Weigold, M.F., and M. Friedman (2003) 'Cultivating the Science Internet Audience', *Science Communication*, 24 (3), 309–332.

Trumbo, C.W., Sprecker, K.J., Dumlao, R.J., Yun, G.W., and S. Duke (2001) 'Use of E-mail and the Web by Science Writers', *Science Communication*, 22 (4), 347–378.

Tuchman, G. (1978) *Making News*. New York: Free Press.

Tudelft (2008) Delft University of Technology website http://www.tudelft.nl/live/pagina.jsp?id=083bdbc5-e2a6-403a-a57f-afef767d3f07&lang=en accessed 25 March 2008.

Turner, L. (2003) 'The Tyranny of "Genetics" ', *Nature Biotechnology*, 21 (11), 1282.

Turney, J. (2002) 'Understanding and Engagement, the Changing face of Science and Society', *Wellcome News*, Q3 (32), 6–7.

van Amerom, M. and M. Ruivenkamp (2006) 'Image Dynamics in Nanotechnology's Risk Debate', Paper presented at Second International Seville Seminar on 'Future-Oriented Technology Analysis: Impact of FTA Approaches on Policy and Decision-Making', Seville 28th–29th September.

van Merkerk, R.O. and D.K.R. Robinson (2006) 'Characterizing the Emergence of a Technological Field: Expectations, Agendas and Networks in Lab-on-a-Chip Technologies', *Technology Analysis & Strategic Management*, 18 (3/4), 411–428.

Wahlberg, A. and L. Sjoberg (2000) 'Risk Perception and the Media', *Journal of Risk Research*, 3 (1), 31–50.

Waldron, A.M., Spencer, D., and C.A. Batt (2006) 'The Current State of Public Understanding of Nanotechnology', *Journal of Nanoparticle Research*, 8 (5), 569–575.

Walsh, B. (2007) *Environmentally Beneficial Nanotechnologies: Barriers and Opportunities*. London: DEFRA.

Weigold, M.F. and D. Treise (2004) 'Attracting Teen Surfers to Science Web Sites', *Public Understanding of Science*, 13 (3), 229–248.

Whitehouse, D. (2007) 'Science Reporting's Dark Secret', *The Independent*, 23 July.

Wilcox, S.A. (2003) 'Cultural Context and the Conventions of Science Journalism: Drama and Contradiction in Media Coverage of Biological Ideas about Sexuality', *Critical Studies in Media Communication*, 20 (3), 225–247.

Wilkinson, C., Allan, S., Anderson, A., and A. Petersen (2007) 'From Uncertainty to Risk?: Scientific and News Media Portrayals of Nanoparticle Safety', *Health, Risk and Society*, 9 (2), 145–157.

Williams, C., Kitzinger, J., and L. Henderson (2003) 'Envisaging the Embryo in Stem Cell Research: Rhetorical Strategies and Media Reporting of the Ethical Debate', *Sociology of Health and Illness*, 25 (7), 793–814.

Wilsdon, J., Wynne, B., and J. Stilgoe (2005) *The Public Value of Science or How to Ensure Science Really Matters*. London: Demos.

Wood, S., Geldart, A., and R. Jones (2008) 'Crystallizing the Nanotechnology Debate', *Technology Analysis and Strategic Management*, 20 (1), 13–27.

Wood, S., Jones, R., and A. Geldart (2003) *The Social and Economic Challenges of Nanotechnology*. London: Economic and Social Research Council.

Wood, S., Jones, R., and A. Geldart (2007) *Nanotechnology, from the Science to the Social: The Social, Ethical and Economic Aspects of the Debate*. Swindon: Economic and Social Research Council. http://www.esrcsocietytoday.ac.uk/ESRCInfoCentre/Images/ESRC_Nano07_tcm6-18918.pdf accessed 16 May 2008.

Woodrow Wilson Project on Emerging Technology (2007) 'A Nanotechnology Consumer Project Inventory'. Woodrow Wilson: Washington, D.C. http://www.nanotechproject.org/inventories/consumer/accessed 16 May 2008.

Wynne, B. (1995) 'Public Understanding of Science', in S. Jasanoff, G.E. Markle, J.C. Petersen and T. Pinch (eds) *Handbook of Science and Technology Studies*. Thousand Oaks: Sage.

Wynne, B. (2006) 'Public Engagement as a Means of Restoring Public Trust in Science – Hitting the Notes, but Missing the Music', *Community Genetics*, 9 (3), 211–220.

Young, P. (1997) 'Writing Articles from Science Journals', in D. Blum and M. Knudson (eds) *A Field Guide for Science Writers*. New York: Oxford University Press.

Index